高等职业教育教材

Chemical Analysis

化学分析

（中英对照版）

夏德强　主编

毛建梅　夏海军　副主编

化学工业出版社

·北京·

内容简介

本书是顺应"走出去"企业人才培养的需要和推动现代职业教育高质量发展的要求而编写，体系完整、内容新颖、插图清新、中英文对照、数字资源丰富，既富有普通高等教育的学科特点，又突出了职业教育的类型特征，具有较强的创新性、先进性、实用性。全书内容主要包括定量分析误差、有效数字及其运算、滴定分析法等，重点介绍酸碱滴定、配位滴定、氧化还原滴定、沉淀滴定的基本原理和基本操作。

本书既可作为本专科职业院校资源环境、生物化工、食品药品等相关学科或专业的双语教材，也能作为普通本科学校相关专业教材，以及相关"走出去"企业员工本地化培养或"引进来"企业员工的培训教材，还可供其他专业师生及分析检验工作者参考。

图书在版编目（CIP）数据

化学分析/夏德强主编；毛建梅，夏海军副主编．—北京：化学工业出版社，2022.9

ISBN 978-7-122-41787-9

Ⅰ.①化⋯　Ⅱ.①夏⋯②毛⋯③夏⋯　Ⅲ.①化学分析-高等学校-教材　Ⅳ.①O65

中国版本图书馆CIP数据核字（2022）第112643号

责任编辑：刘心怡
责任校对：赵懿桐
装帧设计：关　飞

出版发行：化学工业出版社
　　　　　（北京市东城区青年湖南街13号　邮政编码100011）
印　　刷：北京云浩印刷有限责任公司
装　　订：三河市振勇印装有限公司

787mm×1092mm　1/16　印张 $14\frac{1}{4}$　字数 377 千字
2022 年 11 月北京第 1 版第 1 次印刷

购书咨询：010-64518888
售后服务：010-64518899
网　　址：http：//www.cip.com.cn
凡购买本书，如有缺损质量问题，本社销售中心负责调换。

定　　价：46.00 元　　　　　　　　　　　版权所有　违者必究

前言

习近平总书记2013年提出了"一带一路"（The Belt and Road，缩写B&R）倡议。"十三五"期间，中国石油和化工企业"走出去"步伐不断加速，市场份额进一步增大，在全球产业链、供应链、价值链中发挥了重要作用。然而，"走出去"企业尤其是石油化工企业的员工本土化培养培训迫切需要相关的教学资源。2021年10月中共中央办公厅、国务院办公厅印发《关于推动现代职业教育高质量发展的意见》，明确提出要通过提升中外合作办学水平、拓展中外合作交流平台、推动职业教育走出去等措施打造中国特色职业教育品牌，推出一批具有国际影响力的专业标准、课程标准、教学资源。基于此，为顺应"走出去"企业人才培养的需要和国家职业教育改革发展的要求，本教材编写组以新修订实施的《中华人民共和国职业教育法》为依据，以科学性、先进性、适用性为目标，突出职业教育类型特色，紧扣实际、力求实用，解构重构了《分析化学》内容，编写了《化学分析（中英对照版）》和《仪器分析（中英对照版）》。

《化学分析（中英对照版）》共分6章。第1章为概论，第2章为滴定分析法，第3～6章分别为酸碱滴定法、配位滴定法、氧化还原滴定法和重量分析法。

本书既可作为职业本科院校及高等职业专科学校资源环境、生物化工、食品药品等相关学科或专业的双语教材，也能作为普通本科学校相关专业教材，以及相关"走出去"企业员工本地化培养、"引进来"企业员工培训的教材，还可供其他专业师生及分析检验工作者参考。

本书第1～3章、第6章的中文内容及部分英文内容由兰州石化职业技术大学夏德强教授编写，第4章和第5章中文内容由安徽省宣城市环境监测中心夏海军高级工程师编写。本书英文部分的前言、第1章、第2章、第5章、第6章由兰州石化职业技术大学毛建梅翻译；第3章、第4章由兰州石化职业技术大学徐静老师翻译。兰州石化职业技术大学郑晓明、于娇娇、汪永丽、代学玉、柳樱华、张雅迪等老师参与了本书有关章节的审核校对、实验实训项目选题及课后习题的编写工作。全书由夏德强统稿并担任主编，毛建梅、夏海军担任副主编，兰州石化职业技术大学冷宝林教授任主审。

本书的编写参考了大量的相关教材、专著、论文、规范及标准等，尤其是书中部分插图引用了相关的国外教材，在此对本书所引用成果的单位和个人表示衷心感谢。

由于编者的知识和能力水平有限，书中不足之处在所难免，恳请广大师生、读者、专家批评指正，以便今后进一步修订。

夏德强
2022年4月

PREFACE

In 2013, General Secretary Xi Jinping proposed the *"The Belt and Road"* initiative, which was actively responded by petroleum and chemical enterprises in China. During the period of the 13th Five-Year Plan, these companies accelerated the pace of globalization and kept enlarging their market share, so they played a crucial part in industrial chain, supply chain and value chain all over the world. However, the staff in the "global-oriented" enterprises, especially for those who work in petrochemical industry, urgently needs relevant teaching materials for training. Moreover, in October 2021, General Office of the CPC Central Committee and the General Office of the State Council issued the document entitled *Opinions on Promoting the High-quality Development of Modern Vocational Education* where it was clearly stated that a brand of vocational education with Chinese characteristics should be created by promoting Sino-foreign cooperation level in running schools, expanding exchange platforms between China and other countries as well as urging vocational education to go abroad, and then a number of professional curriculum standards and teaching resources with great international influence should be made. In view of the above-mentioned, in order to meet the needs of enterprise talent training and the requirements of the national vocational education reform and development, followed by the *Vocational Education Law of the People's Republic of China*, writers and editors compiled the textbook of *Chemical Analysis* (Chinese-English Edition) & *Instrumental Analysis* (Chinese-English Edition) with the aims of highlighting vocational education's features and emphasizing pragmatic under the guide of scientificity, advancement and applicability.

Chemical Analysis (Chinese-English Edition) is split up into 6 chapters. Chapter 1 and 2 are overviews of analytical chemistry and titrimetric analysis respectively; Acid-base titration, coordination titration, redox titration and gravimetric analysis are discussed from chapter 3 to 6 separately. This textbook can be used as a training material both for staff in local petrochemical industry with globalized intention and for employees of foreign enterprises in China. It can also be used as bilingual textbook for biology and chemical engineering, environmental protection and other related majors in Vocational Institutions for Undergraduates, Higher Vocational Colleges and Normal Undergraduate Universities.

The Chinese of the preface, chapter 1-3, chapter 6 in the textbook were written by Professor Xia Deqiang from Lanzhou Petrochemical University of Vocational Technology. Xia Haijun, Senior Engineer of Xuancheng Environmental Monitoring Center, was responsible for Chinese of chapters 4 and 5. Preface, chapter 1,2,5,6 were translated by Mao Jianmei. Chapters 3 and 4 were translated by Xu Jing. Xia Deqiang also served as the final editor, compiler and editor-in-chief of the book; Mao Jianmei and Xia Haijun served as deputy editors. Leng Baolin, Zheng Xiaoming, Yu Jiaojiao, Wang Yongli, Dai Xueyu, Liu Yinghua and Zhang Yadi participated in the proofreading selection of experimental training projects and compilation of after-class exercises.

The compilation of this coursebook takes a large number of related textbooks, papers, specifications and standards for reference, and I would like to extend my sincere and heartfelt thanks to the copyholders for generous permission to make use of the pieces here.

Suggestions for improvement will be gratefully received.

Xia Deqiang
2022.04

CONTENTS

Chapter 1　Introduction　/ 001

Section 1　Classification of Analytical Chemistry　/ 001
Section 2　Errors in Quantitative Analysis　/ 004
Section 3　Significant Figures and Its Arithmetic Rules　/ 010
Section 4　General Steps of Chemical Analysis　/ 016

Chapter 2　Titrimetric Analysis　/ 021

Section 1　Introduction　/ 021
Section 2　Concentration Expression and Preparation of Standard Solution　/ 025
Section 3　Calculation of Analytical Result　/ 028
Section 4　Main Analytical Apparatus in Titrimetric Analysis　/ 030

Chapter 3　Acid-base Titration　/ 053

Section 1　Theoretical Grounding for Acid-base Balance　/ 054
Section 2　Distribution of Acid-base Species in Aqueous Solution　/ 061
Section 3　pH Calculations for Acid-base Solutions　/ 066
Section 4　Buffer Solution　/ 070
Section 5　Acid-base Indicators　/ 073
Section 6　Titrations of Monoprotic Acid-base　/ 076
Section 7　Preparation and Calibration of Acid-base Standard Solution　/ 087

Chapter 4　Coordination Titration　/ 099

Section 1　EDTA and Its Complexes　/ 099
Section 2　Coordination Dissociation Equilibrium and Influencing Factors　/ 104
Section 3　Principle of Complexometric Titration　/ 110
Section 4　Metallochromic Indicators　/ 115

Section 5 Methods to Improve the Selectivity of Complexometric Titration / 121
Section 6 Application of Complexometric Titration / 126

Chapter 5 Redox Titration / 135

Section 1 Redox Reaction / 135
Section 2 Redox Titration / 142
Section 3 Permanganate Titration / 149
Section 4 Dichromate Titration / 155
Section 5 Iodometric Method / 158

Chapter 6 Gravimetric Analysis and Precipitation Titration / 169

Section 1 Introduction of Gravimetric Analysis / 169
Section 2 Volatilization Gravimetry / 170
Section 3 Precipitation Gravimetry / 172
Section 4 Calculation of Analytical Result / 187
Section 5 Precipitation Titrations / 190

Appendix / 213

Reference / 219

Chapter 1　Introduction
第1章　概　论

 Study Guide　学习指南

Analytical chemistry is a very important science and one of the fastest-growing disciplines in recent years. Analytical chemistry relies on experimental data to draw conclusions, which is an experiment-based science. In this chapter, we shall focus on classification of analytical chemistry, characterization and calculation methods of errors in quantitative analysis, and effective numbers and its arithmetic rules.

分析化学是一门很重要的科学，也是近年来发展最为迅速的学科之一。分析化学依靠实验数据得出结论，是一门以实验为基础的科学。本章重点介绍分析化学的分类，掌握定量分析中误差的表示方法和计算方法，掌握有效数字及其运算规则。

Section 1　Classification of Analytical Chemistry
第1节　分析化学的分类

Analytical Chemistry, rich in content, can be divided into qualitative analysis, quantitative analysis and structural analysis according to different tasks. **Qualitative analysis** is to determine what elements or ions compose the matter as well as the functional group and molecular structure for the organic matter, while **quantitative analysis** is to measure the content of various components of the matter. **Structural analysis** is to study molecular structure, crystal structure or comprehensive form of a substance. In addition, it can be divided into the following categories according to the object analyzed, measurement principle, the sample amount, measured component content and requirements of production department.

分析化学的内容十分丰富，按任务分为定性分析、定量分析和结构分析，定性分析的任务是鉴定物质由哪些元素或离子所组成，对于有机物质还需要确定其官能团及分子结构；定量分析的任务是测定物质各组成部分的含量；结构分析的任务是研究物质的分子结构、晶体结构或综合形态。除此之外，还可根据分析对象、测定原理、试样用量、被测组分含量多少和生产部门的要求，分为如下不同类别。

1. Chemical Analysis and Instrumental Analysis

Chemical analysis is an analytic method on a basis of the material chemical reaction, including titrimetric analysis and gravimetric analysis, which is mainly used for the determination of high content and medium content components. Physical and physical chemistry analysis is an analytic method on a basis of matter physics and the physicochemical properties. This kind of method all needs the special instruments, so it usually is called **instrumental analysis**, which includes absorptiometry, electrochemical analytical methods, chromatography, mass spectrography and radio analytical methods etc. A wide range of types, moreover, some new analysis methods are increasingly appearing (fig.1-1).

1. 化学分析和仪器分析

以物质的化学反应为基础的分析方法称为化学分析法，主要有滴定分析法和重量分析法，主要用于高含量和中含量组分的测定。以物质的物理和物理化学性质为基础的分析方法称为物理分析法和物理化学分析法，这类方法都需要特殊的仪器，因而通常称为仪器分析法。仪器分析法主要有光学分析法、电化学分析法、色谱分析法、质谱分析法和放射化学分析法等，种类很多，而且新的分析方法正在不断出现（图1-1）。

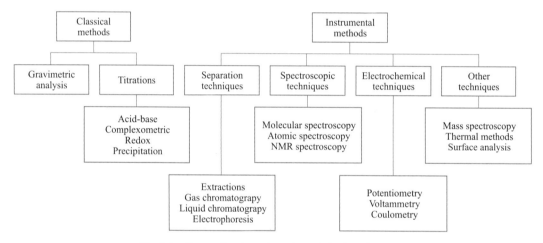

Fig.1-1 General categories of analytical techniques

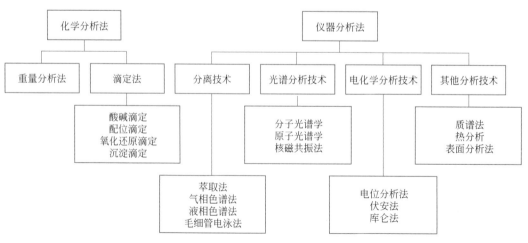

图 1-1 分析方法分类

2. Inorganic Analysis and Organic Analysis

The object of inorganic analysis is inorganic compounds but that of organic analysis is organic compounds. As inorganic compounds contain a wide range of elements, inorganic analysis is adopted to determine what elements, ions, atomic groups or compounds usually compose the sample and the content of various components. In contrast, organic compounds do not contain many elements but have more than 10 million types owing to their complex structures, therefore organic analysis not only includes element analysis but also functional group analysis and structure analysis.

3. Constant Analysis, Semi-micro Analysis and Micro Analysis

In the chemical analysis, according to the amount of sample, it is classified into the constant analysis, the semi-micro analysis and the micro analysis, as is shown in table 1-1. In addition, according to the range of measured components, it can be divided into constant components (>1%), micro components (1% ～ 0.01%) and trace components (0.01%) analysis.

2. 无机分析和有机分析

无机分析的对象是无机化合物，有机分析的对象是有机化合物。在无机分析中，无机化合物所含的元素种类繁多，通常要求鉴定试样是由哪些元素、离子、原子团或化合物所组成，各组分的含量是多少。在有机分析中，虽然组成有机化合物的元素种类不多，但由于有机化合物结构复杂，其种类已达千万种以上，故分析方法不仅有元素分析，还有官能团分析和结构分析。

3. 常量分析、半微量分析和微量分析

分析工作中根据试样用量的多少可分为常量分析、半微量分析和微量分析，如表 1-1 所示。另外，按被测组分含量范围又可分为：常量组分（>1%）、微量组分（1% ～ 0.01%）和痕量组分（<0.01%）分析。

Table 1-1 Analytic method based on the amount of sample
表 1-1 根据试样量分类的分析方法名称

Analysis method name 分析方法名称	Constant analysis 常量分析	Semi-micro analysis 半微量分析	Micro analysis 微量分析
Quality of solid sample /g 固态试样质量 /g	1 ～ 0.1	0.1 ～ 0.01	< 0.01
Volume of liquid sample /mL 液态试样体积 /mL	10 ～ 1	1 ～ 0.01	< 0.01

4. Routine Analysis, Fast Analysis and Arbitrary Analysis

Routine analysis refers to the inspection and control analysis performed by the analysis laboratory on the quality of raw materials and products in the daily production process, which is also called General Analysis.

4. 例行分析、快速分析和仲裁分析

例行分析是指一般分析实验室对日常生产流程中的原材料和产品质量所进行检验控制的分析，又叫"常规分析"。

Fast analysis mainly provides information for controlling production process. For example, analysis in front of stove in the steel mill, it needs to show the analysis result as soon as possible in order to control production process, quick speed is a top priority, the accuracy is the second, it only meets certain requirements.

Arbitrary analysis refers that authoritative organization is allowed to accurately analyze the result by the established standard methods to make sure accuracy of original analysis results, different determination results are obtained from the different departments for the same sample analysis so the dispute happens. Obviously, in arbitrary analysis, the higher accuracy is needed for analytic method and results.

快速分析主要为控制生产过程提供信息。例如，炼钢厂的炉前分析，要求在尽量短的时间内报出分析结果，以便控制生产过程，这种分析要求速度快，准确度达到一定要求便可。

仲裁分析是因为不同的单位对同一试样分析得出不同的测定结果，并由此发生争议时，要求权威机构用公认的标准方法进行准确的分析，以裁判原分析结果的准确性。显然，在仲裁分析中，要求分析方法和分析结果有较高的准确度。

Section 2　Errors in Quantitative Analysis
第 2 节　定量分析的误差

The purpose of quantitative analysis is to accurately determine the content of the component to be tested in the sample through a series of analysis steps. However, in the process of analysis, the results obtained cannot be absolutely accurate because it is limited by some certain subjective factors and objective conditions. Therefore, the next step is to judge the reliability of analysis results and check the causes of errors, so as to take corresponding measures to reduce the errors and make the analysis results as close to the objective truth as possible.

定量分析的目的是通过一系列分析步骤准确测定试样中待测组分的含量。但是，在分析过程中，由于受某些主观因素、客观条件的限制，所得的结果不可能绝对准确。这就需要判断分析结果的可靠程度，查找产生误差的原因，以便采取相应措施减少误差，使分析结果尽量接近客观真实值。

1. Accuracy and Precision

1. 准确度与精密度

Accuracy and precision
准确度与精密度的关系

Accuracy is the proximity between the analysis results and true values. The smaller the difference between them lies, the higher the accuracy of analysis results is. During the actual analysis, in order to obtain reliable analysis results, people always measure several parallel samples under the same conditions, and then take average value. If several data results are close, the precision of analysis is high. It is proved that **precision** is the

准确度是指分析结果与真实值接近的程度。它们之间的差值越小，则分析结果的准确度越高。为了获得可靠的分

proximity among several parallel measurement results. How to evaluate the analysis results from precision and accuracy? Figure 1-2 is a schematic diagram showing results of analyzing the content of Fe_2O_3 in the same sample by A, B, C and D. The dotted line at 65.15% in the figure indicates the true value. Therefore, the analysis results from four people can be shown as follows:

A: Accuracy and precision are both high, the results are reliable;

B: Precision is high, but accuracy is low;

C: Precision and accuracy are very low;

D: Although the average value is close to the true value, several data results are significantly different from each other, and only because positive and negative errors offset each other, it coincidentally makes the result close to the true value, the result is also unreliable.

Precision is a prerequisite for ensuring accuracy. Low precision leads to unreliable results, which loses the premise to measure accuracy. High precision does not mean high accuracy. It is possible to find the cause of inaccuracy, and then correct it to make measurement resuit precise and accurate (Fig.1-3).

析结果，在实际分析中，人们总是在相同条件下对试样平行测定几份，然后取平均值，如果几个数据比较接近，说明分析的精密度高。所谓精密度就是几次平行测定结果相互接近的程度。

如何从精密度和准确度两方面评价分析结果呢？图1-2是A、B、C、D四人分析同一铁矿石试样中Fe_2O_3含量的结果示意图。图中65.15%处的虚线表示真实值，由此，可评价四人的分析结果如下：A所得结果准确度与精密度均好，结果可靠；B的精密度虽高，但准确度较低；C的精密度与准确度均很差；D的平均值虽也接近于真实值，但几个数据彼此相差甚远，而仅是由于正负误差相互抵消才凑巧使结果接近真实值，因而其结果也是不可靠的。

精密度是保证准确度的先决条件。精密度差，所测结果不可靠，就失去了衡量准确度的前提。但高的精密度不一定能保证高的准确度，可以找出精密而不准确的原因，而后加以校正，就可以使测定结果既精密又准确（见图1-3）。

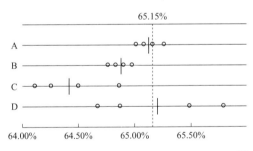

Fig.1-2 The results of analyzing the same sample from different people
（○ means individual measured value, | means average）

图1-2 不同人员分析同一试样的结果
（○表示个别测量值，|表示平均值）

Low accuracy, low precision　　Low accuracy, high precision　　High accuracy, low precision　　High accuracy, high precision

Fig.1-3 The relationship between accuracy and precision

图1-3 精密度与准确度的关系

2. Error and Deviation

(1) Error——a measure of accuracy There are two ways to express error: absolute error (E) and relative error (E_r). The absolute error refers to the difference between the measured value (x_i) and the true value (μ), shown in formula (1.1), while the relative error is the percentage of the absolute error equivalent to the true value, which is shown in formula (1.2).

$$E = x_i - \mu \tag{1.1}$$

$$E_r = \frac{E}{\mu} \times 100\% = \frac{x_i - \mu}{\mu} \times 100\% \tag{1.2}$$

(2) Deviation——a measure of precision The true value is not known in the actual analysis. In view of this, the arithmetic average value of multiple parallel determinations [eq.(1.3)] is usually used to express the analysis result. The difference between each measured value and the average is named deviation. Deviation is divided into absolute deviation [eq.(1.4)] and relative deviation [eq.(1.5)]. The smaller the deviation is, the higher the precision of the measurement obtains.

$$\bar{x} = \frac{x_1 + x_2 + \cdots + x_n}{n} = \frac{1}{n}\sum_{i=1}^{n} x_i \tag{1.3}$$

$$d_i = x_i - \bar{x} \tag{1.4}$$

$$d_r = \frac{d_i}{\bar{x}} \times 100\% \tag{1.5}$$

3. Systematic Error and Random Error

In figure 1-2, why does the result from B has a high precision but low accuracy? Why are there large or small differences for parallel measurement results every time? It is because that there exist various types errors in the process of analysis. Errors can be divided into two categories according to their features, such as systematic error and random error.

(1) Systematic error This kind of error is caused by some special reason. It has unidirectionality, that is, positive and negative, the size has a certain regularity. Systematic errors reappear when the measurement is repeated. If the

2. 误差与偏差

（1）误差——衡量准确度的高低 误差有两种表示方法：绝对误差和相对误差。绝对误差是测量值（x_i）与真实值（μ）之间的差值，即式（1.1）；相对误差是绝对误差相对于真实值的百分比，即式（1.2）。

（2）偏差——衡量精密度的高低 在实际分析工作中，真值并不知道，一般采取多次平行测定的算术平均值［式（1.3）］来表示分析结果。各测定值与平均值之差称为偏差。偏差分为绝对偏差［式（1.4）］和相对偏差［式（1.5）］。偏差越小说明测定的精密度越高。

3. 系统误差与随机误差

在图1-2的例中，为什么B的结果精密度很好而准确度差呢？为什么每人所做的四次平行测定结果都有或大或小的差别呢？这是由于在分析过程中存在着各种性质不同的误差。误差按性质不同可分为两类：系统误差和随机误差。

（1）系误统差 这类误差由某种固定的原因造成的，它具有单向性，即正负、大小都有一定的规律性。当重复进行测定时，系统误差会重复出现。若能

cause can be found and corrected, the systematic error can be eliminated, so it is also called measurable error. The results from B (in fig.1-2) are of high precision and poor accuracy due to systematic errors. The main causes for systematic errors are as follows.

① **Method error** It refers to the error caused by the analysis methods itself. For example, in the titration analysis, the titration end point determined by the indicator does not completely coincide with the chemometric point and side reactions occur, which make measurement result higher or lower completely.

② **Instrument error** It is mainly caused by the instrument itself, inaccurate or not calibrated. For example, when the balance, scale and vessel have errors, the measurement results will be inaccurate in the process of working.

③ **Reagent error** It is error caused by impure reagent or trace impurities in distilled water.

④ **Operation error** It is caused by unreasonable operation for the operator. For example, some people are sensible for judgment of the color change of the end point ,but others are dull;the dropper reading is high or low.

(2) **Random error** Random error refers to the error caused by the random variation of various factors for measured result, such as temperature, humidity, pressure fluctuations, small changes for the instrumental performance, slight deviations in operation, which will cause the analysis results to fluctuate in a certain range, errors happen. Since the random error depends on a series of random factors in the measurement process, whose magnitude and direction are not fixed, it cannot be measured and corrected. Therefore, the random error is also called indeterminate error.

Random error inevitably happens, which objectively exists, and it is difficult to detect and control, there seems to be no regularity from its appearance, but after eliminating the systematic error and measuring it for many times under the same conditions, it can be found that the distribution of random errors is also regular. Generally speaking, it obeys the statistical discipline of normal distribution.

找出原因，并设法加以校正，系统误差就可以消除，因此也称为可测误差。B（图1-2）所做结果精密度高而准确度差，就是由于存在系统误差。系统误差产生的主要原因如下：

① 方法误差　指分析方法本身所造成的误差。例如滴定分析中，由指示剂确定的滴定终点与化学计量点不完全符合以及副反应的发生等，都将系统地使测定结果偏高或偏低。

② 仪器误差　是指主要是仪器本身不够准确或未经校准所引起的。如天平、砝码和容量器皿不准等，在使用过程中就会使测定结果产生误差。

③ 试剂误差　是指由于试剂不纯或蒸馏水中含有微量杂质引起的误差。

④ 操作误差　是由于操作人员的主观原因造成。例如，对终点颜色变化的判断，有人敏锐，有人迟钝；滴管读数偏高或偏低等。

（2）随机误差　随机误差是指测定值受各种因素的随机变动而引起的误差，如温度、湿度、气压的波动，仪器性能的微小变化，操作稍有出入等，都将使分析结果在一定范围内波动，从而造成误差。由于随机误差取决于测定过程中一系列随机因素，其大小和方向都不固定，因此无法测量，也不可能校正，所以随机误差又称为不可测误差。

随机误差不可避免，客观存在，难以觉察，难以控制，从表面上看似乎没有规律，但是消除系统误差后，在同样条件下多次测定，则可发现随机误差的分布也是有规律的，一般服从正态分布统计规律。

① Positive and negative errors of similar magnitudes have the same chance to appear, that is, errors with similar absolute values and opposite signs occur at the same frequency. Therefore, for the data with equally accurate measurement, its algebra sum of random errors tends to be zero.

② Tiny errors occur more frequently, while large errors occur less frequently.

In statistics, the entire range of the random variable x is called overall values, and a set of measurement values $x_1, x_2, x_3,..., x_n$ are randomly selected from overall values as samples. The above rules can be shown by the normal distribution curve (fig. 1-4). In the figure, μ is the average value from infinite multiple determinations, which is the true value when the system error is corrected. The ordinate of the graph represents the probability of an error, and the abscissa indicates standard deviation by means of σ. The figure shows that the probability of the analysis result ranging from $\mu \pm \sigma$ is 68.3%, its probability ranging from $\mu \pm 2\sigma$ is 95.4%, and its probability ranging from $\mu \pm 3\sigma$ is 99.7%. The analysis result whose errors exceed $\pm 3\sigma$ has a probability of 0.3%. Therefore, through multiple determinations, the method of averaging can reduce the influence of random errors on the measurement results.

In addition to systematic and random errors, there exists

① 大小相近的正误差和负误差出现的机会相等，即绝对值相近而符号相反的误差是以同等的机会出现的，因而等精度大量测量的数据，其随机误差的代数和有趋于零的趋势。

② 小误差出现的频率较高，而大误差出现的频率较低。

在统计学中，将随机变量 x 取值的全体称为总体，从总体中随机抽取一组测量值 $x_1, x_2, x_3,..., x_n$ 称为样本。上述规律可用正态分布曲线（图1-4）表示。图中 μ 为无限多次测定的平均值，在校正了系统误差的情况下，即为真值。图的纵坐标代表误差发生的概率，横坐标以标准偏差 σ 为单位。由图可知，分析结果落在 $\mu\pm\sigma$ 的概率为68.3%；落在 $\mu\pm2\sigma$ 的概率为95.4%；落在 $\mu\pm3\sigma$ 的概率为99.7%。误差超过 $\pm3\sigma$ 的分析结果出现的概率为0.3%。因此，通过多次测定，取平均值的方法可以减少随机误差对测量结果的影响。

除了系统误差和随机误差外，在

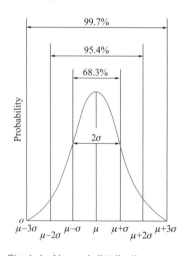

Fig.1-4　Normal distribution curve
图1-4　正态分布曲线

"gross error" due to negligence or incorrect operation in the analysis. For example, wrong reagents, splashed loss test solution, wrong reading scales, incorrect recording, etc., they are all unreasonable faults. As long as the operator has a strong sense of responsibility, being meticulous and obeying strictly operating procedures during the work, faults can be completely avoided. Once a large error occurs, it is determined by analysis that the fault is caused by negligence, it should be discarded when calculating the average value.

4. Calibration and Elimination of Errors

To improve the accuracy of the analysis results, various types of errors mentioned above can be corrected by means of control experiment, blank test, calibration instruments, etc.

(1) **Control experiment**　It is an effective method for testing systematic errors. The known standard sample (or standard solution) with accurate content is being analyzed and tested by the same method to check whether the analysis result is consistent with the standard value. If there is any difference, the calibration data will be found. Control experiment can also be performed by different analytical methods, or by different analysts to test the same experiment.

During production, when the sample is often analyzed, the same method is used to analyze the standard sample to check whether the operation is correct ,whether the instrument is normal. If the result of analysis standard meets the "tolerance", it means that the operation and the instrument meet the requirements, so the results of this analysis are reliable.

(2) **Blank test**　Whose measurement is performed in accordance with the analysis steps and conditions of the sample without reagent a sample, is called Blank Test. Its result is called the "blank value". Deducting the blank value from the analysis results of the sample eliminates systematic errors caused by impurities from reagents, distilled water, laboratory vessels, and the

分析中还会遇到由于过失或差错造成的"过失误差"。例如，加错试剂、试液溅失、读错刻度、记录错误等，这些都属于不应有的过失，只要加强责任感，工作中认真细致、一丝不苟，严格遵守操作规程，过失是完全可以避免的。一旦出现很大的误差，经分析确定是由于过失引起的，则在计算平均值时应舍弃。

4. 误差的校正及消除

以上各类误差可以采用对照试验、空白试验、校准仪器等方法加以校正，以提高分析结果的准确度。

（1）对照试验　这是用来检验系统误差的有效方法。用已知准确含量的标准试样（或标准溶液），按同样方法进行分析测试，检验测试结果与标准值是否一致，如有差异，找出校正数据。对照试验也可以用不同的分析方法，或由不同的分析人员测试同一试验，互相对照。

生产中，常在分析试样的同时，用同样的方法分析标样，以检查操作是否正确、仪器是否正常，若分析标样的结果符合"公差"规定，说明操作与仪器均符合要求，试样的分析结果是可靠的。

（2）空白试验　在不加试样的情况下，按照试样的分析步骤和条件而进行的测定叫做空白试验。得到的结果称为"空白值"。从试样的分析结果中扣除空白值，就可以消除由试剂、蒸馏水、实验器皿和环境带入的杂质所引起的系统误差。空白值过大时，必须采取提纯试

environment. When the blank value is too large, it must be reduced by taking some measures such as purifying reagents or using appropriate vessels.

(3) **Calibration instrument** In the daily analysis, the instrument has been calibrated at the factory, as long as the instrument is properly stored, it is generally not necessary to perform calibration. For the analysis with higher accuracy requirements, the instruments, such as burettes, pipettes, volumetric flasks, balance weights, must be calibrated to obtain the correction value and used it in the calculation of the results to eliminate the error caused by the instrument.

(4) **Calibration method** The systematic error of some analysis methods can be directly calibrated by other methods. For example, in the gravimetric analysis, it is absolutely impossible to precipitate the component to be tested, and other methods must be used to correct the dissolution loss. For example, after precipitating silicic acid, a small amount of silicon remaining in the filtrate can be determined by colorimetry. When the high accuracy is required, the colorimetric measurement result of the component in the filtrate should be added to the gravimetric analysis result.

剂或改用适当器皿等措施来降低。

（3）校准仪器　在日常分析工作中，因仪器出厂时已进行过校正，只要仪器保管妥善，一般可不必进行校准。在准确度要求较高的分析中，对所用的仪器如滴定管、移液管、容量瓶、天平砝码等，必须进行校准，求出校正值，并在计算结果时采用，以消除由仪器带来的误差。

（4）方法校正　某些分析方法的系统误差可用其他方法直接校正。例如，在重量分析中，使被测组分沉淀绝对完全是不可能的，必须采取其他方法对溶解损失进行校正。如在沉淀硅酸后可再用比色法测定残留在滤液中的少量硅，在准确度要求高时，应将滤液中该组分的比色测定结果加到重量分析结果中去。

Section 3　Significant Figures and Its Arithmetic Rules
第3节　有效数字及其运算规则

It is very important for students who begin to learn analytical chemistry to know concept of significant figures, correct readings in the analysis experiments, proper original records, correct processing of the original data, and accurate representation of analysis results.

建立有效数字的概念，在分析实验中正确读数，正确做好原始记录，正确处理原始数据，正确表示分析结果，对于开始学习分析化学的学生十分重要。

1. Significant Figures and Its Digits

1. 有效数字及位数

Significant figures are the figures that can be actually measured by the analytical instrument. Only the last

有效数字是指分析仪器实际能够测量到的数字。在有效数字中只有最

digit in the significant figures is the estimated value, which is suspicious or inaccurate. For example, when weighing on an analytical balance, the absolute error of weighing is ±0.0001 g. If quality of the sample is 0.3280 g, record it as 0.3280 g (4 significant figures), the second "0" is suspicious.

The recorded number, which is usually regarded as the suspicious number, may have an error of ± 1, indicating that the true quality is 0.3279~0.3281 g. The absolute error of weighing at this time is ±0.0001 g, and the relative error is

$$\frac{\pm 0.0001\text{g}}{0.3280\text{g}} \times 100\% = \pm 0.03\%$$

If weighing result is recorded as 0.328g, the actual quality of the sample will be certain value ranging from 0.328g ± 0.001g, that is, the absolute error is ± 0.001g, and the relative error is ± 0.03%.

Therefore, when you write more or fewer "0" at the end of the decimal point during recording, it does not matter much in a mathematical view, but measurement accuracy reflected by the record is virtually exaggerated or reduced by 10 times. The recording figure is not consistent with instrumental accuracy. Each number is important, which represents certain quantity in the data, and it must be noted including the number "0".

The number "0" has a double meaning in the data. If it is an ordinary number, it is significant figures, if it is only used for positioning, it is not significant figures, for example:

1.0002g	5 significant figures
0.5000g, 27.03%, 6.023×10²	4 significant figures
0.0320g, 1.06×10⁻⁵	3 significant figures
0.0074g, 0.30%	2 significant figures
0.6g, 0.007%	1 significant figures

Three "0" in the middle of 1.0002 g, and three "0" in the back of 0.5000 g are all significant figure. The "0" in 0.0074 g is only for positioning, not a significant number.

末一位数字是估计值，是可疑的。例如，在分析天平上称量，称量的绝对误差为 ±0.0001g，假如称得试样的质量为 0.3280g，记录为：0.3280g（四位有效数字），其中最后一个"0"为可疑值。

记录的数字，通常理解为可疑的那位数可能有 ±1 的误差，表示真实质量在 0.3279～0.3281g。此时称量的绝对误差为 ±0.0001g，相对误差为

若将上述称量结果记录为0.328g，则该物体的实际质量为0.328g±0.001g 范围内的某一数值，即绝对误差为 ±0.001g，而相对误差则为 ±0.03%。

可见，记录时在小数点后末尾多写一位或少写一位"0"数字，从数学角度看关系不大，但是记录所反映的测量精确程度无形中被放大或缩小了10倍，这样的记录与测量所用的仪器的准确度是不符合的。所以数据中代表一定量的每一个数字都是重要的，是"0"也得记上。

数字"0"在数据中具有双重意义。若作为普通数字使用，它就是有效数字；若它只起定位作用就不是有效数字，例如：

5 位有效数字
4 位有效数字
3 位有效数字
2 位有效数字
1 位有效数字

在 1.0002g 中间的三个"0"、0.5000g 中后边的三个"0"，都是有效数字；0.0074g 中的"0"只起定位作用，

In 0.0320 g, the front "0" is used for positioning, and the last "0" is a significant figure. Similarly, the last digit of these numbers are indefinite numbers. Therefore, when recording measurement data and calculating results, according to accuracy of measurement instrument used, only the last one of significant figure retained is estimated "uncertain digit".

For some values commonly used in analytical chemistry, the significant figures are as follows:

The quality of the sample
  0.4370g (Analytical balance weighing)
  4 significant figures

Titrant volume
  18.34mL(Burette reading)
  4 significant figures

Reagent volume
  12mL(Measuring cylinder)
  2 significant figures

Standard solution concentration
  0.1000 mol·L^{-1}
  4 significant figures

Tested component content
  23.47%
  4 significant figures

Dissociation constant
  K_a=1.8×10^{-5}
  2 significant figures

Stability Constants of Coordination Compounds
  K_{MgY}=1.00×10$^{8.7}$
  3 significant figures

pH
  4.30, 11.02
  2 significant figures

2. Significant Figures and Calculations

In the process of ordinary analysis and determination, there are several measurement steps, then calculate based on the measured data, and finally obtain the analysis results. However, the measurement accuracy

不是有效数字；在 0.0320g 中，前面的"0"起定位作用，最后一位"0"是有效数字。同样，这些数字的最后一位都是不定数字。因此，在记录测量数据和计算结果时，应根据所使用的测量仪器的准确度，使所保留的有效数字中，只有最后一位是估计的"不定数字"。

分析化学中常用的一些数值，有效数字位数如下：

试样的质量
  0.4370g（分析天平称量）
  4 位有效数字

滴定剂体积
  18.34mL（滴定管读取）
  4 位有效数字

试剂体积
  12mL（量筒量取）
  2 位有效数字

标准溶液浓度
  0.1000 mol·L^{-1}
  4 位有效数字

被测组分含量
  23.47%
  4 位有效数字

解离常数
  K_a=1.8×10^{-5}
  2 位有效数字

配合物稳定常数
  K_{MgY}=1.00×10$^{8.7}$
  3 位有效数字

pH
  4.30, 11.02
  2 位有效数字

2. 有效数据的运算

通常的分析测定过程，往往包括几个测量环节，然后根据测量所得数据进行计算，最后求得分析结果。但是各个测量环节的测量精度不一定完全一致，

in each step may not be completely similar, so the significant figures of several measurement datum may also be different. In the calculation, extra numbers must be rounded off. China's National Standard (GB) has the following rules for rounding.

① Among the number to be removed, if the first number on the left is less than 5 (excluding 5), it is rounded off. For example, if you want to round 14.2432 to three digits, the figure "432" starting from the fourth digit is about to be removed, the first number on the left is "4", less than 5, which should be rounded off, so it is about 14.2.

② Among the number to be removed, if the first number on the left is more than 5 (excluding 5), then add one in the former digit. For example, the figure 26.4843 is rounded to 26.5.

③ Among the numbers to be removed, if the first number on the left is 5, and the numbers on the right are not all zero, then add one in the former digit. For example, the figure 1.0501 is rounded as 1.1.

④ Among the numbers to be removed, when the first number on the left is 5, and the numbers on the right are all zero, if the number to be reserved in the last digit is odd, add one in the former digit, if it is even (including "0"), it doesn't add any figure. For example:

$$0.3500 \rightarrow 0.4$$
$$0.4500 \rightarrow 0.4$$
$$1.0500 \rightarrow 1.0$$

⑤ If the number to be rounded off is more than two figures, it may not be rounded consecutively. For example, the figure 215.4546 is rounded to three significant figures, and it should be 215 at a time. If the figure 215.4546 is rounded to 215.455, 215.46, and then to 215.5, finally to 216, it is incorrect.

3. Calculation Rules of Significant Figures

We now consider how many digits to retain in the answer after you have performed arithmetic operations with your

因而几个测量数据的有效数字位数可能也不相同，在计算中要对多余的数字进行修约。我国的国家标准（GB）对数字修约有如下的规定。

① 在拟舍弃的数字中，若其左边的第一个数字小于5（不包括5）时，则舍去，例如，欲将14.2432修约成三位，则从第4位开始的"432"就是拟舍弃的数字，其左边的第1个数字是"4"，小于5，应舍去，所以修约为14.2。

② 在拟舍弃的数字中，若其左边的第一个数字大于5（不包括5）时，则进一。例如，26.4843 → 26.5。

③ 在拟舍弃的数字中，若其左边的第一个数字等于5，其右边的数字并非全部为零时，则进一。例如，1.0501 → 1.1。

④ 在拟舍弃的数字中，若其左边的第一个数字等于5，其右边的数字皆为零时，所保留的末位数字若为奇数则进一，若为偶数（包括"0"），则不进。例如：

$$12.25 \rightarrow 12.2$$
$$12.35 \rightarrow 12.4$$
$$1225.0 \rightarrow 1.22 \times 10^3$$
$$1235.0 \rightarrow 1.24 \times 10^3$$

⑤ 拟舍去的数字，若为两位以上数字时，不得连续进行多次修约。例如，需将215.4546修约成三位，应一次修约为215。若215.4546 → 215.455 → 215.46 → 215.5 → 216，则是不正确的。

3. 有效数字的运算规则

答案中保留几位有效数字要在数据运算后确定。舍入应该只针对最终答案

data. Rounding should only be done on the final answer (not intermediate results), to avoid accumulating round-off errors. Retain all of the digits for intermediate results in your calculator or spreadsheet.

(1) Addition and subtraction　When several figures are added or subtracted, their sum or difference can only retain a suspicious digit, which should be based on the number with the fewest digits after the decimal point (that is, the largest absolute error). For example, 53.2, 7.45, and 0.66382 are added together. If each figure is recorded in accordance with the significant figures regulations, the last digit is a suspicious number, the "2" in 53.2 is already a suspicious number, so the first decimal is already suspicious after adding the three numbers, which determines the absolute error of the sum. Therefore, the sum of three figures above should not be written as 61.31382, but should be rounded to about 61.3.

If the numbers of digits to be added or subtracted are the same, the answer retains the same decimal digits as any individual numbers. The significant figure in the answer may be more or less than that in the original data.

（1）加减法　当几个数据相加或相减时，它们的和或差只能保留一位可疑数字，应以小数点后位数最少（即绝对误差最大的）的数字为依据。例如，53.2、7.45 和 0.66382 三数相加，若各数据都按有效数字规定所记录，最后一位均为可疑数字，则 53.2 中的"2"已是可疑数字，因此，三数相加后第一位小数已属可疑，它决定了总和的绝对误差。因此，上述数据之和，不应写作 61.31382，而应修约为 61.3。

如果要加减的数字的位数相同，则答案将保留与任何单个数字相同的小数位数。计算答案中的有效数字可能会多于或少于原始数据的位数。

$$\begin{array}{r} 1.362\times 10^{-4} \\ + \ 3.111\times 10^{-4} \\ \hline 4.473\times 10^{-4} \end{array} \qquad \begin{array}{r} 5.345 \\ + \ 6.728 \\ \hline 12.073 \end{array} \qquad \begin{array}{r} 7.26\times 10^{14} \\ - \ 6.69\times 10^{14} \\ \hline 0.57\times 10^{14} \end{array}$$

(2) Multiplication and division　When several figures are multiplied and divided, the retention of significant figures for product or quotient should be based on the figure with the largest relative error, namely, the figure with the least significant figures. For example:

（2）乘除法　几个数据相乘除时，积或商的有效数字位数的保留，应以其中相对误差最大的那个数据，即有效数字位数最少的那个数据为依据。例如：

$$\frac{0.0243\times 7.105\times 70.06}{162.4}=?$$

Because the last digit is a suspicious number, the relative error of each figure is:

因最后一位都是可疑数，各数据的相对误差分别为：

$$\frac{\pm 0.0001}{0.0243}\times 100\%=\pm 0.4\%$$

$$\frac{\pm 0.001}{7.015}\times 100\%=\pm 0.01\%$$

$$\frac{\pm 0.01}{70.06}\times 100\%=\pm 0.01\%$$

$$\frac{\pm 0.1}{164.2} \times 100\% = \pm 0.06\%$$

Obviously, the relative error of 0.0243 is the largest (it is also the figure with the fewest digits), so the result of the above calculation formula should be retained only three significant figures:

$$\frac{0.0243 \times 7.10 \times 70.1}{162} = 0.0747$$

The following rules should be noted, when calculating and rounding the number of significant figures.

① If the first significant figure in a certain figure is more than or equals to 8, the digit of significant figure can be added one. For example, 8.15 can be regarded as four significant figures.

② In analytical chemistry calculations, there are often multiples and fractions, such as 2, 5, 10, 1/5, 1/10, etc. The numbers can be regarded as enough accurate figures regardless of the number of significant figures, while the number of significant figures for calculation results should be determined by other measurement data.

③ In the calculation process, in order to improve the reliability of calculation results, you can temporarily retain more one significant figure, and when the final result is obtained, the extra digit should be removed according to the rules of digital rounding.

④ In analytical chemistry calculations, for the calculation of various chemical equilibrium constants, two or three significant figures are generally reserved. For calculation of various errors, one significant figure is sufficient, and at most two digits. For the calculation of pH, usually only one or two significant figures can be retained, such as pH is 3.4, 7.5, 10.48.

⑤ When drawing a graph on a computer, consider whether the graph is meant to display qualitative behavior of the data or precise values that must be read with several significant figures. If someone will use the graph (figure 1-5) to read points, it should at least have tick marks on both sides of the horizontal and vertical scales. Better still is a fine grid superimposed on the graph.

在计算和取舍有效数字位数时，还要注意以下几点：

① 若某一数据中第一位有效数字大于或等于8，则有效数字的位数可多算一位，如8.15可视为四位有效数字。

② 在分析化学计算中，经常会遇到一些倍数、分数，如2、5、10及$\frac{1}{5}$、$\frac{1}{10}$等，这里的数字可视为足够准确，不考虑其有效数字位数，计算结果的有效数字位数，应由其他测量数据来决定。

③ 在计算过程中，为了提高计算结果的可靠性，可以暂时多保留一位有效数字位数，得到最后结果时，再根据数字修约的规则，弃去多余的数字。

④ 在分析化学计算中，对于各种化学平衡常数的计算，一般保留两位或三位有效数字。对于各种误差的计算，取一位有效数字即已足够，最多取两位。对于pH的计算，通常只取一位或两位有效数字即可，如pH为3.4、7.5、10.48。

⑤ 在计算机上绘制图形时，要考虑该图是为了显示数据的定性行为，还是必须用几个重要的参数读取的精确值。如果使用该图（图1-5）来读取点，那么它至少应该在纵坐标和横坐标的两侧都有标记。最好的方法是将一个固定的网格叠加在图形上。

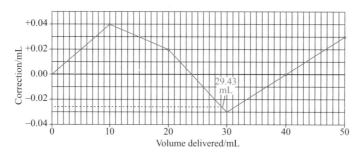

Fig.1-5 Calibration curve for a 50 mL buret [The volume delivered can be read to the nearest 0.1 mL. If your buret reading is 29.43 mL, you can find the correction factor accurately enough by locating 29.4 mL on the graph. The correction factor on the ordinate (y-axis) for 29.4 mL on the abscissa (x-axis) is –0.03 mL (to the nearest 0.01 mL).]

图 1-5 50mL 滴定管的校准曲线（流出体积可以读取到 0.1mL。可以通过校准曲线定位 29.4mL 对应的校正因子，读取滴定管读数为 29.43mL。即横坐标上 29.4mL 对应的纵坐标 –0.03mL 为校正因子。）

Section 4 General Steps of Chemical Analysis

第 4 节 化学分析的一般步骤

The analytical process often begins with a question that is not phrased in terms of a chemical analysis. The question could be "Is this water safe to drink?" or "Does emission testing of automobiles reduce air pollution?" A scientist translates such questions into the need for particular measurements. An analytical chemist then chooses or invents a procedure to carry out those measurements.

When the analysis is complete, the analyst must translate the results into terms that can be understood by others—preferably by the general public. A critical feature of any result is its reliability. What is the statistical uncertainty in reported results? If you took samples in a different manner, would you obtain the same results? Is a tiny amount (a trace) of analyte found in a sample really there or is it contamination from the analytical procedure? Only after we understand the results and their limitations can we draw conclusions. Table 1-2 shows the general steps in the analytical process:

How do you measure the caffeine content of a chocolate bar? Extracting fat from chocolate to leave defatted solid residue for analysis. In this case, fat had to be removed

分析过程通常从一个不是用化学分析术语表述的问题开始。如"喝这种水安全吗？"或者"汽车的排放测试是否减少了空气污染？"科学家会将这些问题转化为对特定测量的需要。然后分析化学家选择或发明一种方法来进行这些测量。

当分析完成后，分析工作者必须将结果转化为其他人可以理解的通用术语。任何分析结果的关键特征就是可靠性。报告结果中的统计不确定性是什么？如果你以不同的方式取样，你会得到相同的结果吗？样品中发现的少量（痕量）分析物真的存在还是分析过程的污染？只有在了解了结果及其局限性之后，我们才能得出结论。表 1-2 展示了分析过程中的一般步骤。

例如，你如何测量巧克力棒中的咖啡因含量？这就需要从巧克力中提取脂肪，留下解冻的固体残留物进行分析。

Table 1-2　The total analytical process
表1-2　分析的一般过程

Formulating the question 提出问题	Translate general questions into specific questions to be answered through chemical measurements 将一般问题转化为需要通过化学分析来解答的具体问题
Selecting analytical procedures 选择分析方法	Search the chemical literature to find appropriate procedures or, if necessary, devise new procedures to make the required measurements 查阅化学文献，找到适当的方法，如果必要，可以设计新的方法以满足分析的需要
Sampling 取样	Sampling is the process of selecting representative material to analyze. If you begin with a poorly chosen sample or if the sample changes between the time it is collected and the time it is analyzed, results are meaningless. "*Garbage in—garbage out*!" 取样就是选择有代表性的试样进行分析的过程。如果您从开始就选择了一个不理想的样本，或者样本在采样或分析过程中发生了变化，那么结果就毫无意义了。正所谓"无用输入，无用输出！"
Sample preparation 样品制备	Converting a representative sample into a form suitable for analysis is called sample preparation, which usually means dissolving the sample. Samples with a low concentration of analyte may need to be concentrated prior to analysis. It may be necessary to remove or mask species that interfere with the chemical analysis (centrifugation and filtration are shown in fig.1-6) 将具有代表性的样品转化为适合分析的形式称为样品制备。样品制备时，通常需要溶解；低浓度的分析样品还需要在分析前进行浓缩；有时还需要对干扰物质进行去除或掩蔽（如图1-6所示的离心法和过滤法）
Analysis 分析	Measure the concentration of analyte in several identical aliquots (portions). The purpose of replicate measurements (repeated measurements) is to assess the variability (uncertainty) in the analysis and to guard against a gross error in the analysis of a single aliquot. 　　The uncertainty of a measurement is as important as the measurement itself because it tells us how reliable the measurement is. If necessary, use different analytical methods on similar samples to show that the choice of analytical method is not biasing the result. You may also wish to construct several different samples to see what variations arise from your sampling and sample preparation procedure 　　分析时需要同时测定多份相同等分试样（部分）中分析物的浓度。重复测量的目的是评估分析中的变异性（不确定性），并避免单次分析时出现的误差。 　　测量的不确定度和测量本身同样重要，它揭示了测量的可靠性。如有必要，对同一样品可采用不同的分析方法，以探索分析方法对测定结果的影响。还可以针对几种不同的样品进行分析，以确定取样和制样过程产生的影响

Reporting and interpretation 分析报告及释义	Deliver a clearly written, complete report of your results, highlighting any limitations that you attach to them. Your report might be written to be read only by a specialist (such as your instructor), or it might be written for a general audience (such as a legislator or newspaper reporter). Be sure the report is appropriate for its intended audience 提供一份清晰的、完整的结果报告，突出显示分析的所有附加信息。结果报告可能只由专家阅读，也可能是为一般读者，请务必确保报告适合其阅读对象
Drawing conclusions 结论	Once a report is written, the analyst might not be involved in what is done with the information, such as modifying the raw material supply for a factory or creating new laws to regulate food additives. The more clearly a report is written, the less likely it is to be misinterpreted by those who use it 一旦编写了报告，分析员可能就不会参与对信息的处理，例如修改工厂的原材料供应或制定新的法律来监管食品添加剂。报告写得越清楚，就越不可能被使用它的人误解

from the chocolate, analytes had to be extracted into water, and residual solid had to be separated from the water. The sample preparation—transforming a sample into a state that is suitable for analysis.

在这种情况下，脂肪必须从巧克力中移除，分析物必须被提取到水中，残余的固体必须从水中分离出来，这个过程就是样品的制备——将样品转化为适合分析的状态。

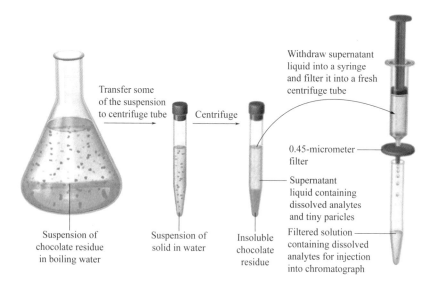

Fig.1-6 Centrifugation and filtration are used to separate undesired solid residue from the aqueous solution of analytes

图1-6 离心法和过滤法常用于分离分析物水溶液中多余的固体残留物

Exercises

1-1 What is the difference between qualitative and quantitative analysis?

1-2 List the general categories of analytical techniques.

1-3 List the steps in a chemical analysis.

1-4 What does it mean to mask an interfering species?

1-5 How many significant figures are there in the following numbers?
   (a) 1.9030
   (b) 0.03910
   (c) 1.40×10^4
   (d) 1200
   (e) 1.0×10^{-5}

1-6 Round each number as requirements:
   (a) 1.2367 to 4 significant figures
   (b) 1.2384 to 4 significant figures
   (c) 0.1352 to 3 significant figures
   (d) 2.051 to 2 significant figures
   (e) 2.0050 to 3 significant figures

1-7 Round each number to three significant figures:
   (a) 0.216 74   (b) 0.216 5   (c) 0.216 500 3

1-8 To deliver a certain volume from a buret, it needs two readings: initial and final. If each reading is made with an uncertainty of ±0.02 mL, what is the relative uncertainty in dispensing 30.00 mL from a buret?

1-9 Calculate the results of the following arithmetic equations with the appropriate number of significant figures. All values are assumed to be experimental.
   (a) $9.388+1.4066-5.2=$
   (b) $4.16 \times 10^3 - 3.79 \times 10^2 =$
   (c) $6.4 \times 1.741 =$
   (d) $7.72 \times 10^{-2} \times 4.116 \times 10^{-3} =$
   (e) $\sqrt{\dfrac{1.5 \times 10^{-8} \times 6.1 \times 10^{-8}}{3.3 \times 10^{-5}}} =$
   (f) $\lg(8.023 \times 10^{-4} + 1.14 \times 10^{-3}) =$

1-10 Find the absolute and percent relative uncertainty and express each answer with a reasonable number of significant figures.
   (a) $9.23 (\pm 0.03) + 4.21 (\pm 0.02) - 3.26 (\pm 0.06) = ?$
   (b) $91.3 (\pm 1.0) \times 40.3 (\pm 0.2) \div 21.1 (\pm 0.2) = ?$
   (c) $[4.97 (\pm 0.05) - 1.86 (\pm 0.01)] \div 21.1 (\pm 0.2) = ?$
   (d) $2.0164 (\pm 0.000\ 8) - 1.233 (\pm 0.002) + 4.61 (\pm 0.01) = ?$
   (e) $2.0164 (\pm 0.000\ 8) \times 10^3 + 1.233 (\pm 0.002) \times 10^2 = ?$
   (f) $[3.14 (\pm 0.05)]^{1/3} = ?$
   (g) $\lg[3.14 (\pm 0.05)] = ?$

1-11 Explain the difference between systematic and random error.

1-12 State whether the errors in (a)-(d) are random or systematic:
   (a) A 25 mL transfer pipet consistently delivers 25.031 mL±0.009 mL.
   (b) A 10 mL buret consistently delivers 1.98 mL±0.01 mL when drained from exactly 0 mL to exactly 2 mL and consistently delivers 2.03 mL±0.02 mL when drained from 2 mL to 4 mL.

(c) A 10 mL buret delivered 1.9839 g of water when drained from exactly 0.00 mL to 2.00 mL. The next time I delivered water from the 0.00 mL to the 2.00 mL mark, the delivered mass was 1.9900 g.

(d) Four consecutive 20.0 μL injections of a solution into a chromatograph were made and the area of a particular peak was 4383, 4410, 4401, and 4390 units.

Chapter 2 Titrimetric Analysis
第 2 章 滴定分析法

 Study Guide 学习指南

Titrimetric analysis is one of the common laboratory methods of chemical analysis that is used to determine the content of an analyte or titrand by adding a standard solution (titrant) into the titrand solution until the quantitative reaction is complete as per stoichiometric relationship. From the required volume and the concentration of the titrant, the amount of the analyte can be calculated. This chapter shall mainly focus on the introduction of the common terminology, methods, characteristics, classification, apparatus, calculation and the preparation of the standard solution. The target of this chapter is to familiarize with the mechanism, analytical procedure and result calculation of the titrimetric analysis methods.

  滴定分析法是经典的化学分析方法之一，它是将一种标准溶液滴加到待测物质的溶液中，直到与待测物质按化学计量关系定量反应为止，然后根据标准溶液的浓度和所消耗的体积，计算待测物质含量的方法。本章重点介绍该方法中的常用名词术语、方法特点、分类、仪器使用、计算以及标准溶液的配制方法等内容。应通过本章的学习，应掌握滴定分析法的原理、测定过程及结果计算。

Section 1 Introduction
第 1 节 概述

1. Concepts and Characteristics of Titrimetric Analysis

1. 滴定分析法概念和特点

(1) Concepts

① **Titrimetric analysis** It is also known as volumetric analysis. It is a common method of chemical analysis that is used to determine the content of an analyte or titrand by adding a standard solution (titrant) into the titrand

（1）滴定分析法概念

① 滴定分析法 又叫容量分析法。这种方法是将一种已知准确浓度的试剂溶液（标准溶液），滴加到被测物质的溶液中，直到所加的试剂与被测物质按

solution until the quantitative reaction is complete as per stoichiometric relationship. From the required volume and the concentration of the titrant, the amount of the analyte can be calculated (Fig.2-1).

The principal requirements for a titration reaction are that it have a large equilibrium constant and proceed rapidly. That is, each increment of titrant should be completely and quickly consumed by analyte until the analyte is used up. Common titrations rely on acidbase, oxidation-reduction, complex formation, or precipitation reactions.

② **Standard solution** (also called titrant) It refers to a reagent of exactly known concentration.

③ **Titrations** It is the process in which the titrant is added to the titrand by means of a burette.

化学计量定量反应为止，然后根据试剂溶液的浓度和用量，计算被测物质的含量（图 2-1）。滴定反应应该具有一个较大的平衡常数，并能快速进行。也就是说，滴定剂的每一个增量都应该被分析物完全和快速地消耗，直到分析物被反应完毕。常见的滴定法有酸碱、氧化还原、配位或沉淀反应。

② 标准溶液（也叫滴定剂） 指一种已知准确浓度的试剂溶液。

③ 滴定 是滴定剂从滴定管加到被测物质溶液中的过程。

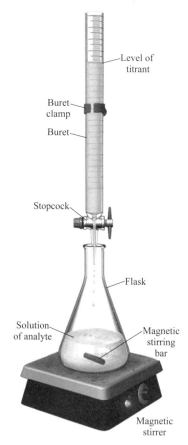

Fig.2-1 Typical setup for a titration (The analyte is contained in the flask, and the titrant is in the buret. The stirring bar is a magnet coated with Teflon, which is inert to most solutions. The bar is spun by a rotating magnet inside the stirring motor.)

图 2-1 典型的滴定装置（待测物质加入到锥形瓶中，滴定剂加入到滴定管中。搅拌子是一种涂有四氟乙烯的磁铁，它对大多数溶液都是惰性的，搅拌子由搅拌电机内的旋转磁铁带动。）

④ **Equivalence point** (also called stoichiometric point) It is the point at which the titrant added just completely reacts with the analyte.

⑤ **Indicator** It refers to a reagent that can indicate the end point of a titration by a color change. An indicator is a compound with a physical property (usually color) that changes abruptly near the equivalence point.

⑥ **Titration end point** the point at which the color of the indicator changes. It is a transition point in the titration at which the color of indicator just changes.

⑦ **Titration end point error** the equivalence point may not always be the same as the end point. The difference between the equivalence point and the end point is called end point error.

(2) Characteristics of titrimetric analysis

① The amount of substance the titrant and the amount of the analyte have stoichiometric relationship.

② This method is suitable for the determination of the analytes with concentration above 1%, sometimes it can also determine trace components.

③ This method is rapid, accurate and easily operated with simple apparatus.

④ This method has wide applications.

2. Requirements of Titration on Chemical Reacion

① The reaction should be complete. If the titration error is required to be within ±0.1%, the reaction should be 99.9% complete when it reaches stoichiometric point.

② There is no side reaction, which is the basis of quantitative calculation.

③ The rate of the reaction is rapid. The reaction completes in a short time as the titrant is added. For slow titration reactions, heating and catalysts can be used to expedite the rate of reaction.

④ There are some methods for determining the

④ 化学计量点 简称"计量点"，当加入的标准溶液与被测物质定量反应完全时，反应达到了"化学计量点"。

⑤ 指示剂 指滴定分析中能指示滴定终点的试剂。当接近滴定终点时，指示剂会有明显的物理性质（通常是颜色）变化。

⑥ 滴定终点 一般依据指示剂的变色来确定。在滴定过程中，指示剂正好发生颜色变化的转变点称为滴定终点。

⑦ 终点误差 滴定终点与计量点不一定恰好符合，由此而造成的分析误差称为"终点误差"。

（2）滴定分析方法特点

① 加入的标准溶液物质的量与被测物质的量恰好是化学计量关系。

② 此法适于组分含量在 1% 以上各种物质的测定，有时也可以测定微量组分。

③ 该法快速、准确、仪器设备简单、操作简便。

④ 用途广泛。

2. 滴定分析对滴定反应的要求

① 完全反应 若要求滴定误差为 ±0.1%，则反应到达化学计量点时完成的程度在 99.9% 以上。

② 无副反应 否则失去定量计算的依据。

③ 反应迅速 加入滴定剂后，反应能立刻完成。某些速度较慢的滴定反应，可通过加热、加催化剂来加快反应速度。

④ 有适当的方法确定终点 能利

end point of the reaction by means of indicators or apparatus. Methods for determining when the analyte has been consumed include detecting a sudden change in the voltage or current between a pair of electrodes, observing an indicator color change, and monitoring absorption of light.

⑤ The co-existed substances do not interfere with the determination.

3. Types of Titration

(1) According the types of reactions between standard solutions and analytes, titrimetric analysis can be classified into:

① Acid-base titration

$$H^+ + OH^- \longrightarrow H_2O$$

② Complexometric titration

$$Zn^{2+} + H_2Y^{2-} \longrightarrow ZnY^{2-} + 2H^+$$

③ Redox titration

$$Cr_2O_7^{2-} + 6Fe^{2+} + 14H^+ \longrightarrow 2Cr^{3+} + 6Fe^{3+} + 7H_2O$$

④ Precipitation titration

$$Ag^+ + Cl^- \longrightarrow AgCl\downarrow$$

(2) According to the methods of titration   Titrimetric analysis can be classified into:

① **Direct titration**   Reactions that fulfill the above five requirements. That is, titrate a analyte directly with a standard solution.

② **Back titration**   When the reaction between the analyte and the titrant is very slow, there is no proper indicator, or the analyte is in a non-soluble solid, a known excess of standard solution is added to the solution to react with the analyte or the solid. When the reaction is complete, titrate the residue with another standard solution. This titration method is called back titration. For example:

$$CaCO_3(s) + 2HCl(excess，过量) \longrightarrow CaCl_2 + H_2O + CO_2\uparrow$$
$$NaOH + HCl(residue，剩余) \longrightarrow NaCl + H_2O$$

用指示剂或仪器分析方法，确定反应的理论终点，如检测一对电极之间电压或电流的突然变化、观察指示剂颜色的变化以及检测光的吸收等。

⑤ 共存物不干扰测定。

3. 滴定分析分类

（1）根据标准溶液与待测物质间反应类型分 滴定分析法分为：

① 酸碱滴定

② 络合滴定

③ 氧化还原滴定

④ 沉淀滴定

（2）按滴定方式分

① 直接滴定法 满足上述 5 个条件的反应，即可用标准溶液直接滴定待测物质。

② 返滴定法 当试液中待测物质与滴定剂反应很慢、无合适指示剂、用滴定剂直接滴定固体试样反应不能立即完成时，可先准确地加入过量标准溶液，与试液中的待测物质或固体试样进行反应，待反应完全后，再用另一种标准溶液滴定剩余的标准溶液，这种滴定方式称为返滴定法。例如：

③ **Replacement titration** If the reaction does not follow a certain equation or there are side reactions, react the analyte with a reagent and replace it quantitatively into another substance. Titrate this substance with a standard solution. This procedure is called replacement titration. For example:

③ 置换滴定法 当待测组分所参与的反应不按一定反应式进行或伴有副反应时，可先用适当试剂与待测组分反应，使其定量地置换为另一种物质，再用标准溶液滴定这种物质，这种滴定方法称为置换滴定。例如：

$$Cr_2O_7^{2-} + 6I^- + 14H^+ \longrightarrow 2Cr^{3+} + 3I_2 + 7H_2O$$

$$\underset{\text{blue} \atop \text{蓝色}}{I_2} + 2S_2O_3^{2-} \longrightarrow \underset{\text{colorless} \atop \text{无色}}{2I^-} + S_4O_6^{2-}$$

④ **Indirect titration** If the analyte does not react with the titrant directly, it sometimes can be determined by another reaction. For example, the determination of Ca^{2+} can be done by the determination of $C_2O_4^{2-}$ with $KMnO_4$.

④ 间接滴定法 不能与滴定剂直接起反应的物质，有时可以通过另外的化学反应，以滴定法间接进行滴定。例如：Ca^{2+} 的测定，用 $KMnO_4$ 测定 $C_2O_4^{2-}$。

$$Ca^{2+} + C_2O_4^{2-} \longrightarrow CaC_2O_4 + H_2SO_4 \longrightarrow CaSO_4 + H_2C_2O_4$$

$$\underset{\substack{\text{Analyte} \\ \text{colorless} \\ \text{试样，无色}}}{5HO-\overset{\overset{O}{\|}}{C}-\overset{\overset{O}{\|}}{C}-OH} + \underset{\substack{\text{Titrant} \\ \text{purple} \\ \text{滴定剂，紫色}}}{2MnO_4^-} + 6H^+ \longrightarrow \underset{\substack{\text{colorless} \\ \text{无色}}}{10CO_2} + \underset{\substack{\text{colorless} \\ \text{无色}}}{2Mn^{2+}} + 8H_2O$$

Section 2 Concentration Expression and Preparation of Standard Solution

第 2 节 标准溶液浓度表示方法及配制

1. Expression of Concentration of Standard Solution

A standard solution is a solution with exactly known concentration. The concentration can be expressed in the following ways.

(1) Molar concentration Molar concentration means the amount of substance the solute in the unit volume solution.

1. 标准溶液浓度表示方法

标准溶液是一种已知准确浓度的溶液，它的浓度表示方法有以下几种。

（1）物质的量浓度 物质的量浓度指单位体积溶液中所含溶质的物质的量。

$$\text{Molar concentration(mol} \cdot \text{L}^{-1}) = \frac{\text{Molar of solute(mol)}}{\text{Volume of solution(L)}} \tag{2.1}$$

Since the value of the amount of substance depends on the choice of the basic unit, therefore the unit must be indicated in the expression. The choice of unit is on the basis of chemical reactions. For instance:

$$c(H_2SO_4) = 0.1 \text{mol} \cdot L^{-1}$$
$$c(2H_2SO_4) = 0.05 \text{mol} \cdot L^{-1}$$
$$c(1/2 H_2SO_4) = 0.2 \text{mol} \cdot L^{-1}$$

(2) Titer

① T_s　Mass of titrant in 1mL standard solution (g), s stands for titrant.

② $T_{x/s}$　Mass of analyte chemically equivalent to 1mL standard solution (g), in which x stands for analyte.

2. Preparation of Standard Solution

In titrimetric analysis, the analytical result depends on the concentration and volume of the standard solution. The correct preparation and standardization of the standard solution concerns the accuracy of analytical results.

(1) Direct method　A certain amount of a primary standard carefully weighed on an analytical balance is dissolved and transfer to a volumetric flask. Then calculate the exact concentration of the solution. For instance, the preparation of $K_2Cr_2O_7$ solution.

Primary standard: a compound that can be used directly to prepare and standardize the standard solution (table 2-1). A primary standard should meet the following requirements:

① Constant composition: the actual composition accords with the chemical formula.

② High purity: >99.9%.

（2）滴定度

① T_s　每 mL 标准溶液含有标准物质的质量（g），s 表示标准物质。

② $T_{x/s}$　每 mL 标准溶液相当于被测物质的质量（g），x 表示被测物质。

2. 标准溶液的配制

在滴定分析中，所得的分析结果是由标准溶液的浓度和其体积决定的，如何准确的配制标准溶液和标定标准溶液，关系到分析结果的准确。

（1）直接法　用分析天平准确称取一定量的基准物质，溶解后，转移到容量瓶中定容，然后算出该溶液的准确浓度。如 $K_2Cr_2O_7$ 的配制。

Preparation of potassium dichromate
重铬酸钾标准溶液的制备

基准物质：能用于直接配制或标定标准溶液的物质（见表2-1）。基准物质须满足：

① 组成恒定：实际组成与化学式符合；

② 纯度高：一般纯度应在99.9%以上；

Table 2-1　Commonly used primary standard
表 2-1　常用基准物质

Name 名称	Chemical formula 分子式	Drying conditions before use 使用前干燥条件
sodium carbonate 碳酸钠	Na_2CO_3	Dry for 2-2.5h at 270-300℃
potassium biphthalate 邻苯二甲酸氢钾	$KHC_8H_4O_4$	Dry for 1-2h at 110-120℃
potassium dichromate 重铬酸钾	$K_2Cr_2O_7$	Dry for 3-4h at 140-150℃
sodium oxalate 草酸钠	$Na_2C_2O_4$	Dry for 1-1.5h at 130-140℃
zinc oxide 氧化锌	ZnO	Dry for 2-3h at 800-900℃
sodium chloride 氯化钠	$NaCl$	Dry for 40-45min at 500-650℃
silver nitrate 硝酸银	$AgNO_3$	Dry to constant weight in concentrated sulfuric acid desiccator

③ Stable properties: no decomposition, hygroscopic, aging or oxidation during storing or weighing.

④ Large molar mass: the larger the weighing mass, the less the weighing error.

⑤ Solubility to water, dilute acid or dilute alkali under application conditions.

(2) Standardization　Many substances cannot be used directly in the preparation of standard solution because their concentrations are instable, hygroscopic, easy to volatile, easy to absorb CO_2 or other impurities in the air. However, these substances can be firstly prepared into a solution with the concentration close to what is need. Then standardize the concentration of the solution with a primary standard, e. g. HCl, NaOH.

③ 性质稳定：保存或称量过程中不分解、不吸湿、不风化、不易被氧化等；

④ 具有较大的摩尔质量：称取量大，称量误差小；

⑤ 使用条件下易溶于水（或稀酸、稀碱）。

（2）标定法　很多物质不能直接用来配制标准溶液，物质本身的浓度不固定、易挥发、易潮解、易吸湿、易吸收空气中的 CO_2 或含有其他杂质。可将其先配制成一种近似于所需浓度的溶液，然后用基准物质来标定它的准确浓度，如 HCl、NaOH 等。

Preparation of NaOH
氢氧化钠溶液配制

① Coarse preparation of solution: weigh a certain amount of substance or a certain volume of solution to prepare into a solution with the concentration close to what is need.

② Standardization: The exact concentration of an indirectly prepared solution must be measured by a primary standard or another standard solution. This process is called standardization.

Procedure of standardization:

Exact weighing (analytical balance) of primary standard → dissolution → indicator → titration with the desired solution to the end point.

① 粗配溶液：粗略称取一定量的物质或量取一定体积溶液，配制成接近于所需浓度的溶液。

② 标定：间接法配制的溶液其准确浓度须用基准物质或另一标准溶液来测定。这种确定其浓度的操作，称为标定。

标定的过程：

准确称取（分析天平）一定量的基准物质→溶解→指示剂→用所需标定的溶液滴定到终点。

3. Notes on Preparation of Solution

① Prepare solutions with pure water. Rinse the container at least three times.

② The solutions should be contained in the reagent bottles. The light sensitive solutions should be contained in amber-coloured bottles. The solutions prepared with volatile reagents should be secured with airtight lids in place. The solutions deteriorating in the air or emitting corrosive gaseous substances should also be secured with airtight lids or sealed with wax if to be stored for a long time. Concentrated alkali should be contained in plastic bottles. If stored in glass bottles, rubber plugs should be used.

③ Each bottle of reagent should be labeled with name, specification, concentration and preparation date.

3. 配制溶液注意事项

① 所用的溶液应用纯水配制，容器应用纯水洗 3 次以上。

② 溶液用带塞的试剂瓶盛装；见光易分解的溶液要装在棕色瓶中；挥发性试剂配制的溶液，瓶塞要严密；见空气易变质及放出腐蚀性气体的溶液要盖紧，长期存放时要蜡封；浓碱液用塑料瓶装，如装在玻璃瓶中，要用橡胶塞塞紧。

③ 每瓶试剂溶液必须标明名称、规格、浓度和配置日期的标签。

Section 3 Calculation of Analytical Result
第 3 节 滴定分析法的结果计算

1. Physical Quantities Commonly Used in Calculation

(1) Amount of substance (n)   A physical quantity

1. 计算中常用的物理量

（1）物质的量（n） 表示物质多少的一个物理量，单位为 mol（摩尔）。

indicating the amount of a substance with unit mol. A mole (mol) has 6.02×10^{23} (Avogadro's number) particles, which can be molecule, atom, ion, electron or other particles or combination of these particles. In analytical chemistry, the basic unit is determined on the basis of chemical reaction.

(2) Molar mass (M)　The mass of substance per mole (g·mol^{-1}). Basic unit should be indicated.

(3) Molarity (c)　the amount of substance per volume solution (mol·L^{-1}, M). A liter (L) is the volume of a cube that is 10 cm on each edge. Because 10 cm=0.1 m, 1 L=(0.1 m)^3=10^{-3} m^3. Chemical concentrations, denoted with square brackets, are usually expressed in moles per liter (c). Thus "[H$^+$]" means "the concentration of H$^+$."

(4) Weight percent(wt%)　The amount of analyte in sample, expressed by percentage or mg·g^{-1}.

The percentage of a component in a mixture or solution is usually expressed as a weight percent (wt%):

$$\text{Weight percent} = \frac{\text{mass of solute}}{\text{mass of total solution or mixture}} \times 100\% \qquad (2.2)$$

Ethanol (CH$_3$CH$_2$OH) is often purchased as a 95% (wt%) solution containing 95 g of ethanol per 100 g of total solution. The remainder is water.

(5) Volume percent (vol%)　Volume percent (vol%) is defined as:

$$\text{Volume percent} = \frac{\text{volume of solute}}{\text{volume of total solution}} \times 100\% \qquad (2.3)$$

(6) Concentration (ρ)　The mass of substance per volume solution (g·L^{-1}, mg·L^{-1}).

2. Calculation Basis in Titrimetric Analysis

If A is analyte, B is standard solution, the titrimetric reaction is:

1mol 指含有 6.02×10^{23}（阿伏伽德罗常数）个的基本单元。基本单元可以是分子、原子、离子、电子及其他粒子，或是这些粒子的特定组合。分析化学中是根据化学反应确定基本单元的。

（2）摩尔质量（M）　表示每摩尔物质的质量（g·mol^{-1}），必须指明基本单元。

（3）物质的量浓度（c）　表示单位体积溶液中所含物质的量（mol·L^{-1}），可用方括号表示，因此 [H$^+$] 的意思是 H$^+$ 的浓度。

（4）质量分数（wt%）　表示待测组分在样品中的含量，可以是百分数或 mg·g^{-1}。

混合物或溶液中某一组分的含量通常用质量分数（wt%）表示：

市售工业乙醇通常是质量分数为 95% 的溶液，也就是每 100g 溶液中含 95g 乙醇。

（5）体积分数（vol%）　体积分数定义：

（6）质量浓度（ρ）　表示单位体积中某种物质的质量，单位可以是 g·L^{-1}、mg·L^{-1} 等。

2. 滴定分析法计算依据

设 A 为待测组分，B 为标准溶液，滴定反应为：

$$aA + bB \longrightarrow cC + dD$$

When A and B react completely as per stoichiometric relationship:

当 A 与 B 按化学计量关系完全反应时，则：

$$\frac{n_A}{n_B} = \frac{a}{b} \tag{2.4}$$

① If the volume V_A of analyte, concentration c_B and volume V_B of standard solution B are known, the concentration of analyte A c_A will be:

① 若已知待测溶液的体积 V_A 和标准溶液的浓度 c_B 和体积 V_B，则待测组分 A 浓度 c_A 为：

$$c_A V_A = \frac{a}{b} \times c_B V_B$$

$$c_A = \frac{a}{b} \cdot \frac{V_B}{V_A} c_B \tag{2.5}$$

② To solve the mass m_A of analyte:

② 求试样中待测组分的质量 m_A：

$$n_A = \frac{a}{b} \cdot n_B$$

$$\frac{m_A}{M_A} = \frac{a}{b} \cdot n_B = \frac{a}{b} \cdot c_B \cdot V_B \times \frac{1}{1000}$$

$$m_A = \frac{a}{b} \cdot c_B V_B M_A \times 10^{-3} \quad (\text{the unit of } V \text{ is mL, } V \text{ 的单位为 mL}) \tag{2.6}$$

③ To solve the mass fraction w_A of analyte in the sample:

③ 求试样中待测组分的质量分数 w_A：

$$w_A = \frac{m_A}{m_s} = \frac{\frac{a}{b} \cdot c_B \cdot V_B \cdot M_A}{m_s} \times 10^{-3} \tag{2.7}$$

m_s is the mass of the sample.

m_s 是试样质量。

Section 4 Main Analytical Apparatus in Titrimetric Analysis
第 4 节 滴定分析法的主要仪器

1. Analytical Balance

1. 分析天平

An electronic balance uses electromagnetic force compensation to balance the load on the pan. Figure 2-2

电子天平使用电磁力补偿来平衡负载。图 2-2 是一种典型的电子天平，

shows a typical analytical balance with a capacity of 100-200 g and a readability of 0.01-0.1 mg. **Readability** is the smallest increment of mass that can be indicated. A microbalance weighs milligram quantities with a readability of 1μg. Common types of balances are shown in table 2-2.

To weigh a chemical, first place a clean receiving vessel on the balance pan. The mass of the empty vessel is called the tare. On most balances, you can press the "TAR" button to reset the tare to 0. Add the chemical to the vessel and read its mass. If there is no automatic tare

其量程为 100～200g，最小分度值为 0.01～0.1mg。分度值是指能够显示的最小质量。微量天平为毫克级，分度值为 1μg。常见天平的种类见表 2-2。

要称量一种化学物质，首先应该在天平托盘上放置一个干净的容器。容器的质量为皮重。对大多数天平而言，可以按"TAR"按钮将皮重重置为 0，然后将该化学物质加入到容器中并读取其

Fig.2-2 Electronic analytical balance measures mass down to 0.1 mg

图 2-2 称量精度为 0.1mg 的电子分析天平

Table 2-2 Common types of balances

表 2-2 常见天平的种类

Analytical Balance	Capacity/g	Readability/g	Resolution
Macroanalytical Balance	50-400	0.0001	$(0.5-4) \times 10^6$
Semimicro balance	30-200	0.00001	$(0.3-2) \times 10^7$
Micro balance	3-20	0.000001	$(0.3-2) \times 10^7$
Ultramicro balance	2	0.0000001	2×10^7
Precision Balances	Capacity	Readability/g	Resolution
Industrial precision scale	30-6000kg	0.0001-0.1kg	$(0.1-6) \times 10^5$
Precision balance	100-30000g	0.001-1g	$(0.3-2) \times 10^5$

*Maximum load: largest mass capacity; Readability: smallest division; Resolution: cap/readability (should be high value).

operation, subtract the tare mass from that of the filled vessel. To protect the balance from corrosion, **chemicals should never be placed directly on the weighing pan**. Also, be careful not to spill chemicals into the mechanism below the balance pan.

An alternative procedure, called **weighing by difference**, is necessary for hygroscopic reagents, which rapidly absorb moisture from the air. First weigh a capped bottle containing dry reagent. Then quickly pour some reagent from the weighing bottle into a receiver. Cap the weighing bottle and weigh it again. The difference is the mass of reagent delivered from the weighing bottle. With an electronic balance, set the initial mass of the weighing bottle to zero with the tare button. Then deliver reagent from the bottle and reweigh the bottle. The negative reading on the balance is the mass of reagent delivered from the bottle (fig.2-3).

质量即可。如果没有自动去皮操作，则应该从装入物质后容器的重量中减去皮重。为了防止天平不受腐蚀，化学物质绝不能直接放在托盘上。此外，还要注意不能将化学物质泄漏到托盘下面的部件中。

还有一种称量方法，称为减量法，其适用于易吸水物质的称量，因为这类物质可以快速吸收空气中的水分。称量时，用滤纸条取出干燥器中的称量瓶，在接收器的上方倾斜瓶身，用瓶盖轻击瓶口使试样缓缓落入接收器中。当估计试样接近所需量时，继续用瓶盖轻击瓶口，同时将瓶身缓缓竖直，用瓶盖向内轻击瓶口，使粘于瓶口的试样落入瓶中，盖好瓶盖。将称量瓶放入天平，显示的质量差即为试样质量（图2-3）。

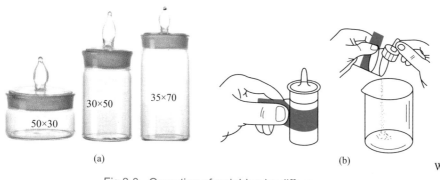

Fig.2-3　Operation of weighing by difference
图2-3　减量法的操作

Weighing liquid sample by difference
减量法称量液体样品

2. Buret

Burets can be classified into acid buret and alkali buret. The acid buret shall be taken as an example in this section. The acid buret in figure 2-4 is a precisely manufactured glass tube with graduations enabling you to measure the volume of liquid delivered through the stopcock (the valve) at the bottom. The 0-mL mark is near the top. If the initial liquid level is 0.83 mL and the final level is 27.16 mL, then you have delivered 27.16−0.83=26.33 (mL). Class A burets (the most accurate grade) are

2. 滴定管

Burets
认识滴定管

滴定管分酸式滴定管和碱式滴定管，本书以酸式滴定管的使用为例。图2-4中的酸式滴定管是一个精确制造的玻璃管，其刻度可以测量通过底部旋塞

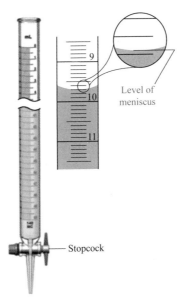

Fig.2-4 Glass buret with Teflon stopcock (Enlargement shows meniscus at 9.68 mL. Estimate the reading of any scale to the nearest tenth of a division. This buret has 0.1 mL divisions, so we estimate the reading to the nearest 0.01 mL.)

图 2-4 具有聚四氟乙烯旋塞的玻璃滴定管（放大图显示弯月面下平面在 9.68mL 处。后续任何刻度线的读数，需精确到刻度的十分之一，该滴定管的刻度为 0.1mL，因此估计读数精确到 0.01mL。）

certified to meet tolerances in table 2-3. If the reading of a 50 mL buret is 27.16 mL, the true volume can be anywhere in the range 27.21 mL to 27.11 mL. Never throw away precision in reading an instrument.

(1) Rinsing of acid buret

① Rinse directly with tap water, soap water or washing

（阀）滴出的液体体积。0mL 的标记接近顶部。如果初始液位为 0.83mL，最终液位为 27.16mL，则滴出液体的体积为 27.16−0.83=26.33（mL）。A 级滴定管（最准确的等级）经认证符合表 2-3 中的公差。如果一个 50mL 滴定管的读数是 27.16mL，真实的体积可以在 27.21mL 到 27.11mL 的范围内。永远不要忽略仪器的精度。

（1）酸式滴定管的洗涤

① 无明显油污的滴定管直接用自

Table 2-3 Tolerances of class A burets
表 2-3 A 级滴定管的公差

Buret volume/mL	Smallest graduation/mL	Tolerance/mL
5	0.01	±0.01
10	0.05 or 0.02	±0.02
25	0.1	±0.03
50	0.1	±0.05
100	0.2	±0.10

powder water for the burets without obvious grease. Scouring powder cannot be used in case the inside surface is scratched and the accuracy is influenced.

来水冲洗或用肥皂水或洗衣粉水泡洗，但不能用去污粉洗以免划伤内壁，影响体积的准确测量。

Rinsing of buret
滴定管的洗涤

② If the buret is smeary, rinse with chromic acid. Remove the water inside the buret and close stopcock. Add 10-15 mL of chromic acid wash solution into the buret. Hold the buret with both hands. Turn and rotate the buret so all inside surfaces have come into contact with the wash solution. Place upright the buret, open the stopcock and drain the wash solution into the original beaker.

③ If the buret is heavily smeary, fill the buret with a plenty of wash solution and stay for more than 10 minutes. Or steep the buret with warm wash solution. After the wash solution is drained, flush with tap water and then with distilled water 3-4 times. The inside surfaces of clean buret should be thoroughly and uniformly wet with water without water drops hanging on the wall.

(2) Greasing of acid buret　Lie the buret on the table. Untie the rubber band fixing the plug and take out the plug. Wipe the plug and the barrel with a piece of clean paper or cloth. If the bore is jammed by spent grease, remove the grease with metal brush. If the tip of buret is blocked by grease, fill the buret with water, then dip the tip into hot water to melt the grease. Suddenly open the stopcock to flush the grease away. Apply a thin layer of petroleum jelly or vacuum grease circularly onto the plug (fig.2-5), getting around the places close to the bore in case it is blocked. After greasing, put the plug back into its barrel and turn the plug in the same direction until it looks transparent when observing from the outside. Fasten the plug into barrel with a rubber band to prevent the out-slipping. The greased plug should fit tightly in its barrel and turn easily so there is no leakage in use.

② 有油污不易洗净时，用铬酸洗液洗涤。洗时应将管内的水尽量除去，关闭活塞，倒入 10～15mL 洗液于滴定管中，两手端住滴定管，边转动边向管口倾斜，直至洗液布满全部管壁为止。立起后打开活塞，将洗液放回原瓶中。

③ 油污严重时，需用较多洗液充满滴定管浸泡十几分钟或更长时间，甚至用温热洗液浸泡一段时间。洗液放出后，先用自来水冲洗，再用蒸馏水淋洗 3～4 次，洗净的滴定管其内壁应完全被水均匀地润湿而不挂水珠。

（2）酸式滴定管的涂油　把滴定管平放在桌面上，将固定活塞的橡胶圈取下，再取出活塞，用干净的纸或布将活塞和塞套内壁擦干（如果活塞孔内有旧油垢塞堵，可用金属丝轻轻剔去，如果管尖被油脂堵塞可先用水充满全管，然后将管尖置热水中，使其溶化，突然打开活塞，将其冲走）。用手指蘸少量凡士林（或真空脂）在活塞孔的两头沿圆周涂上薄薄一层（图2-5），在紧靠活塞孔两旁不要涂凡士林，以免堵住活塞孔。涂完，把活塞放回塞套内，向同一方向转动活塞，直到从外面观察时全部透明为止。然后用橡胶圈套住，将活塞固定在塞套内，防止滑出。涂好油的酸式滴定管活塞与塞套应密合不漏水，并且转动要灵活。

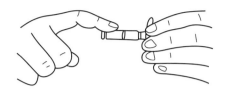

Fig.2-5 Greasing of acid buret

图 2-5 酸式滴定管涂油

Greasing of acid buret
酸式滴定管的涂油

(3) Leakage test of acid buret   Close the stopcock and fill into the buret some distilled water to a certain level. Hold the buret vertically for 2 minutes and observe if the level draws back, or if water leaks from the tip, or if water seeps from the plug. Turn the plug 180° and observe for another 2 minutes. If leakage is found, repeat the procedure in (2).

（3）酸式滴定管的试漏   关闭滴定管活塞，装入蒸馏水至一定刻线，直立滴定管 2min。仔细观察刻线上的液面是否下降，滴定管下端有无水滴漏下，活塞缝隙中有无水渗出。然后将活塞旋转 180°后等待 2min 再观察。如有漏水现象应重新擦干涂油。

Leakage test of buret
滴定管的试漏

(4) Filling of solution and driving of bubbles   Shake well the standard solution in the volumetrice flask to let the water condensated on the inner wall of flask mix into solution. Rinse the buret with 10mL of this standard solution. Open the stopcock and allow one third of the solution to run through to flush the tip, then close the stopcock. Invert the buret to a horizontal position. Turn and rotate the buret so all inside surfaces have come into contact with the solution. Discard this solution from the mouth of buret. Do not open the stopcock in case the grease on the plug is flushed into buret. Empty the buret and rinse for the second time. The tip should be flushed each time. Repeat the operation two to three times so the water residue in the buret can be removed to ensure the constancy of the concentration of solution. Fill the standard solution into the buret directly from the regeant bottle instead of via funnels or other containers. When the buret is filled with standard solution, the tip is still empty. Tilt the buret by 30°, and quickly open the stopcock with left hand to let some solution flush out so the tip of buret is also filled with the solution. When all the air bubbles are out (fig.2-6), fill in

（4）酸式滴定管的装填溶液和赶气泡   将瓶中标准溶液摇匀，使凝结在瓶内壁的水混入溶液，再用该标液润洗滴定管，每次用约 10mL，从下口放出约 1/3 以洗涤尖嘴部分，然后关闭活塞横持滴定管并慢慢转动，使溶液与管内壁处处接触，最后将溶液从管口倒出，但不要打开活塞，以防活塞上的油脂冲入管内，尽量倒空后再洗第二次，每次都要冲洗尖嘴部分，如此洗 2~3 次，即可除去滴定管内残留的水分，确保标准溶液浓度不变。装液时要直接从试剂瓶注入滴定管，不要再经过漏斗等其他容器。当标准溶液装入滴定管时，出口管还没有充满溶液，此时将酸式滴定管倾斜约 30°，左手迅速打开活塞使溶液冲出，就能充满全部出口管，气泡排除后（图 2-6），加入标准溶液至"0"刻度以上，等待 30s，再转动活塞，把液面调节在 0.00mL 刻度处。

Chapter 2   Titrimetric Analysis   035

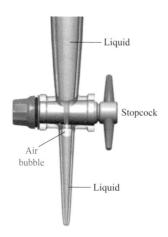

Fig.2-6　An air bubble trapped in the acid buret

图 2-6　酸式滴定管中的气泡

Filling of solution and driving of bubbles
滴定管装填溶液和赶气泡

Adjusting the level to zero
滴定管的调零

the standard solution to zero graduation and wait three seconds. Turn the plug, and adjust the level to zero graduation.

(5) Titration with buret　It's better to carry out the titration in a conical flask, or in a beaker if necessary. During titration, titrate with left hand and shake flask with right hand. The operation of a buret is as follows: the thumb of left hand is in the front, while the forefinger and middle finger are behind the buret pipe. All the fingers are slightly bending inwards to hold the stopcock enclosed in the hollow of the hand to prevent the loosening of plug or the protruding of plug, which might result in the leakage of solution. Hold the neck of Erlenmeyer flask with the thumb, forefinger and the middle finger of the right hand, and shake the flask circularly in the same direction. Do not shake it backwards and forwards. In titration, open the stopcock and control the flow rate of the solution. It is required to be able to: ① add drop-by-drop, ② add only one drop, ③ pend one drop at the tip, which means to learn the technique of adding a partial drop (Fig.2-7). Place the tip of buret into flask one to two centimeters below the neck and titrate at the rate of three or four drops per minute. Do not let the solution flow as a liquid column. Shake while titrate. When the titration is close to the end point, add the solution dropwise or partially. Use a wash bottle to rinse the sides of the flask to be sure all titrant

（5）酸式滴定管的滴定　滴定最好在锥形瓶中进行，必要时也可在烧杯中进行。滴定操作是左手进行滴定，右手摇瓶。使用酸式滴定管的操作为：左手的拇指在管前，食指和中指在管后，手指略弯曲，轻轻向内扣住活塞（图2-7）。手心空握，以免活塞松动或可能顶出活塞使溶液从活塞隙缝中渗出，右手的拇指、食指和中指拿住锥形瓶颈，沿同一方向按圆周摇动锥瓶，不要前后振动。滴定时旋转活塞，控制溶液流出速度，要求做到能：①逐滴放出；②只放出1滴；③使溶液呈悬而未滴的状态，即练习加半滴溶液的技术。滴定时，滴定管尖嘴部分插入锥形瓶口下约1～2cm处，以每秒3～4滴的速度滴定，切不可呈液柱流下。边滴边摇，接近终点时，应1滴或半滴地加入，并用洗瓶吹入少量水冲洗锥形瓶内壁使附着的溶液全部流下，然后摇动锥形瓶，观察终点是否达到，如终点未到，继续滴定，直至准确到达终点（颜色变化半分钟不消失）为止。

Fig.2-7 Use of acid buret

图 2-7 酸式滴定管的使用

Titration with buret
滴定管的滴定

Adding a partial drop
滴定管半滴操作

Controlling the flow rate of the solution
滴定管三种滴定速度控制

is mixed in the flask. Shake the flask well and observe if the end point is reached. If not, keep on titrating until the arrival of end point (the indicator colour change does not disappear in half a minute).

(6) Readings of acid buret When reading the liquid level in a buret, your eye should be at the same height as the top of the liquid. If your eye is too high, the liquid seems to be higher than it really is. If your eye is too low, the liquid appears too low. The error that occurs when your eye is not at the same height as the liquid is called parallax.

（6）酸式滴定管的读数 滴定管读数时，视线应与液面顶端平齐。若视线太高，则读数偏高；若视线太低，则读数偏低。当视线和液面不平齐时，会产生误差，该误差称为视差。

Calibration of buret
滴定管校准

① After the solution is filled or drained, wait thirty seconds or one minute before reading to allow the solution on the inside surfaces to flow down.

② Hold the upper end (without marking) of the buret with thumb and forefinger and maintain its upright position before reading.

③ For colourless solutions and light-coloured solutions, the reading should be the lowest point of the meniscus. This point should be at eye level for reading, i. e. the line of sight should be parallel with the mark. For coloured solutions, the line of sight should be parallel with two highest points of liquid level at sides. The initial reading and the

① 注入溶液或放出溶液后，需等待 30s～1min 后才能读数（使附着在内壁上的溶液流下）。

② 滴定管应用拇指和食指拿住滴定管的上端（无刻线处）使管身保持垂直后读数。

③ 对于无色溶液或浅色溶液，应读弯月面下缘实线的最低点。为此，读数时视线应与弯月面下缘实线的最低点相切，即视线与弯月面下缘实线的最低点在同一水平面上。对于有色溶液，应使视线与液面两侧的最高点相切。初读

final reading should follow the same criterion (fig.2-8)

④ The buret that are back-lined with a blue line follow different criterion in reading for colourless solutions. There are two meniscuses intersecting at a point on the blue line. The line of sight should be parallel with this point (fig.2-9). For coloured solutions, the reading criterion is the same.

⑤ Note the reading on the buret from 0.00mL each time and be accurate to two decimal places 0.01 mL.

⑥ To help read, a reading chart can be placed at the back of the buret with the black part 1mm lower than the bottom of meniscus. Then read the lowest point of the meniscus.

(7) Notes on use of acid buret

① When the titration is complete, the solution residue in the buret should be discarded instead of being drained into the original flask. Rinse the buret with tap water and distilled water for several times, then clamp it upside down to the buret stand.

② If the acid buret is not in use for a long time, the plug should be lined with a piece of paper so that it is easy opening in the future use.

(8) Autotitrator Autotitrator (fig.2-10) delivers reagent from a reservoir bottle into a beaker containing analyte. The electrode immersed in the beaker monitors pH or concentrations of specific ions. Volume and pH readings can be exported to a spreadsheet.

和终读应用同一标准（图2-8）。

④ 有一种蓝线衬背的滴定管，它的读数方法（对无色溶液）与上述不同，无色溶液有两个弯月面交于滴定管蓝线的某一点，读数时视线与此点在同一水平面上（图2-9），对有色溶液读数方法与上述普通滴定管相同。

⑤ 滴定时，要求每次都从0.00mL开始，读数必须准确到0.01mL。

⑥ 为了协助读数，可将读数卡放在滴定管背后，使黑色部分在弯月面下约1mm处，然后读此黑色弯月面下缘的最低点。

（7）酸式滴定管的使用注意事项

① 滴定结束后，滴定管内剩余的溶液应弃去，不要倒回原瓶中。然后依次用自来水、蒸馏水冲洗数次，倒立夹在滴定管架上。

② 酸式滴定管长期不用时，活塞部分应垫上纸。否则时间一久，塞子不易打开。

（8）自动滴定仪 自动滴定仪（图2-10）将滴定剂从储液瓶输送到装有待测液的烧杯中，用浸入烧杯中的电极测定特定离子的浓度或pH值，输送的体积和检测值可以导出到电子表格中。

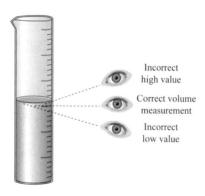

Fig.2-8 Reading of normal buret
图2-8 滴定管的读数

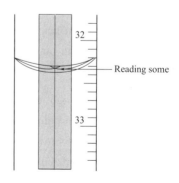

Fig.2-9 Reading of buret with blue lines
图2-9 蓝线滴定管的读数

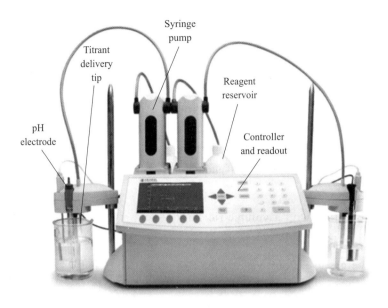

Fig.2-10 Autotitrator
图 2-10 自动滴定仪

3. Volumetric Flask

A volumetric flask (fig.2-11) is used for precise preparation of solution with a definite molar concentration. It is pear-shaped, with a flat bottom and a ground glass stopper. The neck of the volumetric flask is elongated and narrow with graduation mark lines. The mark line indicates the

3. 容量瓶

容量瓶（图 2-11）主要用于准确地配制一定摩尔浓度的溶液。它是一种细长颈、梨形的平底玻璃瓶，配有磨口塞。瓶颈上刻有标线，当瓶内液体在所指定温度下达到标线处时，其体积即为

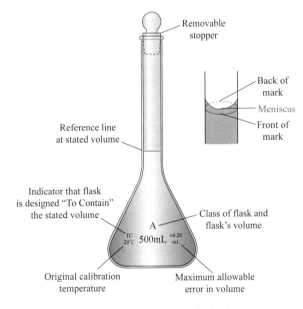

Fig.2-11 Volumetric flask
图 2-11 容量瓶

Chapter 2 Titrimetric Analysis

volume of liquid contained under certain temperature when filled up to that mark. The volumetric flasks are of various sizes, containing 100 mL, 250mL or 500 mL of liquid. A single volumetric flask has only one size. Most volumetric flasks bear the label "TC 20 ℃," which means to contain at 20 ℃. (Pipets and burets are calibrated to deliver, "TD," their indicated volume.) The temperature of the container is relevant because both liquid and glass expand when heated.

(1) Preparation of solution with a volumetric flask (fig.2-12)

① Check leakage before use. Pour water of half of the flask volume into the volumetric flask. Tighten the stopper. Press the stopper of flask with the forefinger of right hand, and support the bottom with the left hand. Place the flask upside down to check if there is leakage. If not, return it to upright position, turn the stopper 180°. Then invert the volumetric flask and check the leakage again. Repeat this process two times to verify that the volumetric flask does not leak round the stopper. Only the volumetric flasks without leakage can be used.

② Place the precisely weighed solid solute in a beaker and dissolve with a small amount of solvent. Transfer the solution into a volumetric flask. Rinse the beaker

瓶上所注明的容积数。一种规格的容量瓶只能量取一个量。常用的容量瓶有100mL、250mL、500mL 等多种规格。大多数烧瓶上都有"TC20℃"的标签，即20℃时的容量。移液管和滴定管都是"TD"（转移）校准的转移装置。由于液体、气体受热膨胀，故容积与温度密切相关。

（1）使用容量瓶配制溶液的方法（图 2-12）

① 使用前检查瓶塞处是否漏水。具体操作方法是：在容量瓶内装入半瓶水，塞紧瓶塞，用右手食指顶住瓶塞，另一只手五指托住容量瓶底，将其倒立（瓶口朝下），观察容量瓶是否漏水。若不漏水，将瓶正立且将瓶塞旋转180°后，再次倒立，检查是否漏水，若两次操作，容量瓶瓶塞周围皆无水漏出，即表明容量瓶不漏水。经检查不漏水的容量瓶才能使用。

② 把准确称量好的固体溶质放在烧杯中，用少量溶剂溶解。然后把溶液转移到容量瓶里。为保证溶质能全部转

Operation of volumetric flask
容量瓶的使用

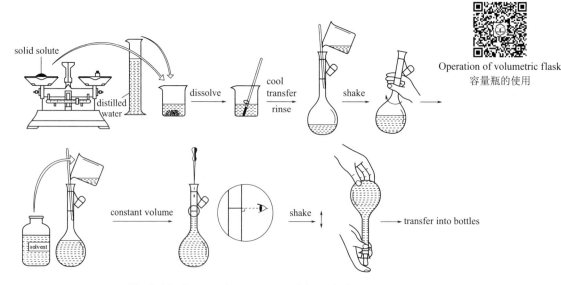

Fig.2-12 Preparation process of the solution
图 2-12 配制溶液

with the solvent several times to ensure that all the solute is transferred. The rinsed solution shall also be transferred to the volumetric flask (fig.2-13). Guide the flow with a glass rod as transferring. That is, leaning one end of the glass rod on the inside surface of the flask neck. Cautions should be taken that the other parts of the glass rod should not touch the flask mouth to avoid flowing of liquid onto the outside surface of the flask.

③ When the liquid level in the volumetric flask is one centimeter to the mark line, carefully add the liquid with a pipette until the bottom of meniscus arrives at the mark. If water is over-added and exceeds the mark line, the preparation has to be re-done.

④ Fasten the stopper and mix well the liquid inside the volumetric flask by inverting and shaking. When the liquid level is still and observed to be lower than the mark, it is because of the loss of a very small amount of liquid on the wetting of flask neck, which can be neglected. Therefore, it is not necessary to fill more water into the flask. Otherwise the concentration of the solution being prepared will drop.

(2) Notes on usage of volumetric flasks

① The volume of volumetric flask is specific, without consecutive graduation marks. Hence a type of volumetric flask can only be used for the preparation of a definite volume of solution. Therefore, find out the volume of the solution to be prepared first and choose a

移到容量瓶中，要用溶剂多次洗涤烧杯，并把洗涤溶液全部转移到容量瓶里（图2-13）。转移时要用玻璃棒引流。方法是将玻璃棒一端靠在容量瓶颈内壁上，注意不要让玻璃棒其他部位触及容量瓶口，防止液体流到容量瓶外壁上。

③ 向容量瓶内加入的液体液面离标线1cm左右时，应改用滴管小心滴加，最后使液体的弯月面与标线正好相切。若加水超过刻度线，则需重新配制。

④ 盖紧瓶塞，用倒转和摇动的方法使瓶内的液体混合均匀。静置后如果发现液面低于刻度线，这是因为容量瓶内极少量溶液在瓶颈处润湿所损耗，所以并不影响所配制溶液的浓度，故不要在瓶内添水，否则，将使所配制的溶液浓度降低。

（2）使用容量瓶时注意事项

① 容量瓶的容积是特定的，刻度不连续，所以一种型号的容量瓶只能配制同一体积的溶液。在配制溶液前，先要弄清楚需要配制的溶液的体积，然后再选用相同规格的容量瓶。

Absolute calibration of volumetric flask
容量瓶的绝对校准

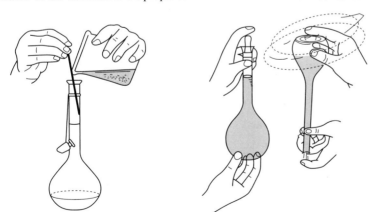

Fig.2-13　Transfer and mix well the solution into a volumetric flask
图 2-13　溶液的转移与摇匀

proper volumetric flask for it.

② The substances that dissolve easily and are not exothermic can be put directly into volumetric flask via a funnel, while the other substances cannot be dissolved in volumetric flasks. These substances should be dissolved in beakers and then transported to volumetric flasks.

③ The quantity of solvent for the washing of beaker cannot surpass the graduation mark of volumetric flask.

④ The volumetric flasks cannot be heated. If a solute is exothermic during dissolution, cool the solution before transferring it to the volumetric flask because the volumetric flasks are generally calibrated at 20 ℃. If a hot solution or a cold solution is poured into a volumetric flask, the flask will expand with heat and contract with cold, resulting in the inaccuracy of the volume measured and the inaccuracy of the solution concentration prepared.

⑤ The volumetric flasks can only be used for the preparation instead of storage of solutions. Solutions might erode the flask bodies and result in the inaccuracy of measurements.

⑥ The volumetric flasks should be washed immediately after use. Plug in the stopper with a piece of paper inserted between the stopper and flask mouth to prevent adhension of them.

4. Transfer pipet and Measuring pipet

A transfer pipet (fig.2-14) is a laboratory tool used to transport a measured volume of liquid. It is a long and narrow glass tube with a bulb in the middle and a mark line at the uppermost position. The bulb is indicated with its volume and the calibration temperature. It comes in a variety of sizes, e. g. 5.00 mL, 10.00 mL and 25.00 mL. Choose proper size of transfer pipet for the delivery of liquid. However, if a small and fractional amount of solution is to be transported, a measuring pipet is generally used.

A measuring pipet (fig. 2-15) is a glass tube with a series of marked lines allowing the delivery of adjustable

② 易溶解且不发热的物质可直接用漏斗倒入容量瓶中溶解，其他物质基本不能在容量瓶里进行溶质的溶解，应将溶质在烧杯中溶解后转移到容量瓶里。

③ 用于洗涤烧杯的溶剂总量不能超过容量瓶的标线。

④ 容量瓶不能进行加热。如果溶质在溶解过程中放热，要待溶液冷却后再进行转移，因为一般的容量瓶是在20℃的温度下标定的，若将温度较高或较低的溶液注入容量瓶，容量瓶则会热胀冷缩，所量体积就会不准确，导致所配制的溶液浓度不准确。

⑤ 容量瓶只能用于配制溶液，不能储存溶液，因为溶液可能会腐蚀瓶体，从而使容量瓶的精度受到影响。

⑥ 容量瓶用毕应及时洗涤干净，塞上瓶塞，并在塞子与瓶口之间夹一纸条，防止瓶塞与瓶口粘连。

4. 移液管、吸量管

Operation of transfer pipet
移液管的使用

移液管（图2-14）是准确移取一定量液体的工具。他是一根细长中间膨大的玻璃管，在管的上端有刻度线。膨大部分标有它的容积和标定时的温度。如需吸取5.00mL、10.00mL、25.00mL等整数，用相应大小的移液管。量取小体积且不是整数时，一般用吸量管。

volume of solution.

Usage of transfer pipet and measuring pipet:

(1) Washing   Before use, transfer pipet and measuring pipet must be washed to the condition that there are no water drops exist on the surfaces. The washing process of pipet is the same as that of buret. First, wash with wash solution first, then flush with tap water and finally clean with distilled water.

(2) Rinsing   To ensure the constant concentration of solution during delivery, dry the inside and outside of the pipet mouth with a piece of filter paper. Rinse the pipet with the solution to be delivered three times to displace the water on the surfaces of pipet. The waste solution should be discarded.

(3) Withdrawal of solution   Hold the pipet with the thumb and the middle finger of right hand on the place above the uppermost marked line. Place the pipet into

吸量管（图 2-15）是带有多刻度的玻璃管，用它可以吸取不同体积的溶液。

使用移液管或吸量管移取溶液的方法：

（1）洗涤　使用前移液管和吸量管都要洗涤，直至内壁不挂水珠为止。方法与洗涤滴定管一样，先用洗液洗，再用自来水冲洗，最后用蒸馏水洗涤干净。

（2）润洗　为保证移取溶液时溶液浓度保持不变，应使用滤纸将管口内外水珠吸去，再用被移溶液润洗三次，置换移液管或吸量管内壁的水分。润洗后的溶液应该弃去。

（3）吸取溶液　吸取溶液时，用右手大拇指和中指拿在管子的刻度上方，插入溶液中，左手用吸耳球将溶液吸入

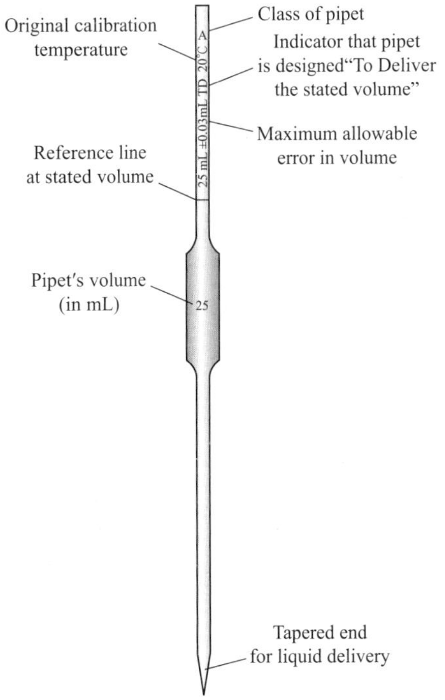

Fig.2-14   Transfer pipet
图 2-14   移液管

Fig.2-15   Measuring pipet
图 2-15   吸量管

the solution and place the pipet rubber suction bulb on the end of the pipet to draw up solution (squeeze the bulb to empty out the air beforehand) as in fig.2-16. The tip of the pipet should be placed at least 1cm beneath the liquid surface. Do not place too deep into solution in case there is too much solution on the outside surface of the pipet. In the meanwhile, do not place too shallow in case air is sucked into the pipet when the liquid level drops. As the level rises above the mark, romove the bulb and press down on the end of the pipet with the forefinger of right hand. In general, thumb is not used for this purpose because it is not as flexible as forefinger. Release pressure from the forefinger so that the liquid level drops slowly until the bottom of the meniscus reaches the mark on the pipet. A better alternative is to use an automatic suction device such as that in figure 2-17 that remains attached to the pipet.

(4) Dispensing of solution   Place the pipet in an beaker flask or a volumetric flask. Tilt the flask so that the tip of pipet leans against the inside surface while the pipet stays upright. It is not correct to let the tip of pipet touch the bottom of flask. Release the forefinger allowing the liquid to flow freely down along the surface of flask. Stay for 15 seconds as all the liquid runs out (if QUICK is indicated on the pipet, do not stay). Take out the pipet.

管中（预先捏扁，排除空气）。吸管下端至少伸入液面1cm，不要伸入太多，以免管口外壁沾附溶液过多，也不要伸入太少，以免液面下降后吸空。用洗耳球慢慢吸取溶液（图2-16），眼睛注意正在上升的液面位置，移液管应随容器中液面下降而降低。当液面上升至标线以上，立即用右手食指按住管口。（一般不用大拇指操作，大拇指操作不灵活。）随后右手食指稍稍抬起，让液面缓慢下降到凹液面与刻度正好相切即可。可以用如图2-17所示的自动吸液仪器替代，这种仪器吸液时与移液管或吸量管连接。

（4）放出整管溶液　将移液管放入锥形瓶或容量瓶中，将锥形瓶或容量瓶略倾斜，管尖靠瓶内壁，移液管垂直。管尖放到瓶底是错误的。松开食指，液体自然沿瓶壁流下，液体全部留出后停留15s（移液管上标有"快"，应该不停留），取出移液管。留在管口的液体不要吹出，因为校正时未将这部分体积

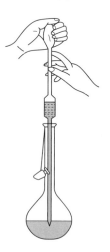

Fig.2-16 Withdrawal of solution
图2-16　吸取溶液

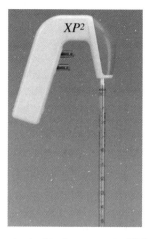

Fig.2-17 Electronic pipet-aid allows you to fill the pipet by pressing the top button and to drain the pipet by pressing the lower button.
图2-17　电子移液管可以通过按下顶部按钮来吸取溶液，并通过按下下部按钮来放出溶液。

Do not blow out the liquid retaining in the mouth of pipet since its volume is not considered in the calibration of the pipet (if BLOW is indicated on the pipet, this amount of liquid should be blown out). If only a certain amount of solution is to be dispensed, the volume of solution is found by calculating the difference of the initial volume and the final volume. Note that the mark should be at eye level. In experiments, use as far as possible the same section of the same pipet to transport liquid (fig.2-18).

Notes: ① Clean the pipet after use and place it on the pipet stand. ② In experiments, pipet and solutions should correspond with each other to avoid contamination.

5. Micropipets

Micropipets (figure 2-19) deliver volumes of 1 to 1 000 μL (1 μL =10^{-6} L). Liquid is contained in the disposable polypropylene tip, which is stable to most aqueous solutions and many organic solvents except chloroform ($CHCl_3$). The tip also is not resistant to concentrated nitric or sulfuric acids. To prevent aerosols from entering the pipet shaft, tips are available with polyethylene filters. Aerosols can corrode mechanical parts of the pipet or cross contaminate biological experiments.

计算在内（移液管上标有"吹"，应该将留在管口的液体吹出）。使用吸量管放出一定量溶液时，通常是液面由某一刻度下降到另一刻度，两刻度之差就是放出的溶液的体积，注意目光与刻度线平齐。实验中应尽可能使用同一吸量管的同一区段的体积（图2-18）。

注意事项：①移液管使用后，应洗净放在移液管架上；②移液管和吸量管在实验中应与溶液一一对应，不应串用以避免沾染。

5. 移液枪

移液枪（图2-19）移取液体的体积为1至1000μL（1μL=10^{-6}L）。液体通过一次性聚丙烯吸头吸入，除氯仿（三氯甲烷）外，吸头对大多数水溶液和许多有机溶剂都是稳定的，但是其同样不能抵抗浓硝酸或硫酸。为了防止气溶胶进入移液管，吸头可以安装聚乙烯过滤器，因为气溶胶会腐蚀移液管的机械部件或造成生物实验交叉污染。

Calibration of transfer pipet
移液管的校正

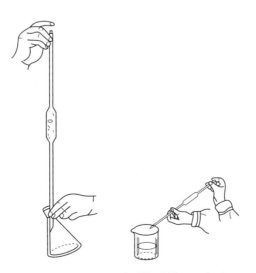

Fig.2-18 Dispensal of liquid from pipet
图2-18 移液管放液

To use a micropipet, place a fresh tip tightly on the barrel. Keep tips in their package or dispenser so that you do not contaminate the tips with your fingers. Set the desired volume with the knob at the top of the pipet. Depress the plunger to the first stop, which corresponds to the selected volume. Hold the pipet vertically, dip it 3-5 mm into the reagent solution, and slowly release the plunger to suck up liquid. Leave the tip in the liquid for a few seconds to allow the aspiration of liquid into the tip to go to completion. Withdraw the pipet vertically from the liquid without touching the tip to the side of the vessel. The volume of liquid taken into the tip depends on the angle at which the pipet is held and how far beneath the liquid surface the tip is held during uptake. To dispense liquid, touch the tip to the wall of the receiver and gently depress the plunger to the first stop. Wait a few seconds to allow liquid to drain down the tip, and then depress the plunger further to squirt out

使用时，首先把一个新的吸头牢牢地固定在吸引管上。注意保持吸头在包装盒或分配器中，以免手指污染吸头。在移液枪的顶部用旋钮设定移液体积。按下活塞到第一停点，对应于设定的移液体积。保持移液枪垂直，将其浸入溶液中 3～5mm，然后缓慢释放柱塞以吸取液体，并将吸头停留在液体中几秒钟，确保液体吸入吸头。垂直将移液枪移出液体，注意不要让吸头接触到容器内壁。吸入吸头的液体体积取决于移液枪的角度以及在吸液过程中吸头在液面下的距离。放液时，吸头紧贴容器壁，轻轻按下柱塞，等待几秒后，让液体排出吸头，然后进一步按下柱塞，排出余液。吸头可以丢弃或用洗瓶冲洗，并重复使用。装有过滤器的移液嘴［图2-19（b）］无法进行清洗并重复使用。

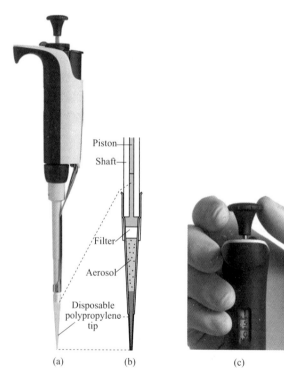

Fig. 2-19 (a) Micropipet with disposable plastic tip; (b) Enlarged view of disposable tip containing polyethylene filter to prevent aerosol from contaminating the shaft of the pipet; (c) Volume selection dial set to 150 μL

图2-19 （a）带有一次性枪头的微量吸管；（b）含有聚乙烯过滤器（防止气溶胶污染移液管）的一次性枪头的放大图；（c）体积选择旋钮设置体积为150μL

the last liquid. The tip can be discarded or rinsed well with a squirt bottle and reused. A tip with a filter [figure 2-19(b)] cannot be cleaned for reuse.

The procedure we just described for aspirating (sucking in) and delivering liquids is called "forward mode." The plunger is depressed to the first stop and liquid is then taken up. To expel liquid, the plunger is depressed beyond the first stop. In "reverse mode," the plunger is depressed beyond the first stop and excess liquid is taken in. To deliver the correct volume, depress the plunger to the first stop and not beyond. Reverse mode with slow operation of the plunger improves precision for foamy solutions (proteins or surfactants) and viscous (syrupy) liquids. Reverse pipetting is also good for volatile liquids such as methanol and hexane. For volatile liquids, pipet rapidly to minimize evaporation.

6. Syringes

Microliter syringes, such as that in figure 2-20, come in sizes from 1 μL to 500 μL and have an accuracy and precision near 1%. When using a syringe, take up and discard several volumes of liquid to wash the glass walls and to remove air bubbles from the barrel. The steel needle is attacked by strong acid and will contaminate strongly acidic solutions with iron. A syringe is more reliable than a micropipet, but the syringe requires more care in handling and cleaning.

Figure 2-21 is an example of a programmable dual syringe diluter that automatically dispenses microliter volumes and can create reproducible mixtures of two solutions from the two syringes.

上述过程为"正向模式"。即先将柱塞压至第一停点，然后吸收液体。为了排出残留的液体，通常要将柱塞压到第二停点。在"反向模式"下，柱塞被压至第二次停点，这时过量的液体被吸入。为移取准确的体积，需要将活塞按到第一停点，且不要超过。反向模式提高了泡沫溶液（蛋白质或表面活性剂）和黏性液体（糖浆）的移取精度。反向移液有利于挥发性液体，如甲醇、六己烷等的移取。对于挥发性液体，移液要快，以减少蒸发。

6. 注射器

微量注射器（如图2-20），规格1μL到500μL，精度和精度接近1%。当使用注射器时，要吸取并排出一定体积的液体清洗内壁，并排出注射器内的气泡。金属针头被强酸腐蚀后会造成溶液污染。注射器比移液枪更可靠，但注射器在使用和清洗时要格外小心。

图2-21是一个可编程双注射器稀释仪，该稀释仪可自动分配微量体积，并可从两个注射器中提取溶液，配成两种溶液的混合物。

Fig.2-20　Hamilton syringe with a volume of 1 μL and divisions of 0.02 μL on the glass barrel
图2-20　容积为1μL、玻璃管刻度为0.02μL的Hamilton注射器

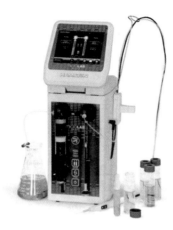

Fig.2-21 Microlab 600 dual syringe diluter
图 2-21 Microlab 600 可编程双注射器稀释仪

Experiment 1 Electronic Analytical Balance and Weighing Exercise

1. Objective

(1) Learn to use analytical balance correctly.
(2) Grasp weighing method: direct weighing; weighing of fixed weigh; weighing by difference.

2. Principle

An electronic balance uses electromagnetic force compensation to balance the load on the pan. In general, it has automatic calibration, zero adjustment, automatic tare, overload indication, fault alarm, automatic weighing results and other functions. Electronic scale shows good features with a long life, stable performance, easy operation and high sensitivity.

Here is a brief description of the use steps of the electronic balance, take as the common BS-224S type of the electronic balance.

Schematic diagram of the BS-224S electronic balance

(1) Turn on: Click the ON button, after a brief self-test, the display screen should show "0.0000 g". If the display screen is not 0.0000 g, press the TAR key.
(2) Weighing: Place lightly things on weighing pan. When the digital on the display screen is to be in stable, you may read and record the weighing results.
(3) Deducting the tare: When a container is placed on the pan, its weight can be displayed. Press

the TAR key, the display appears 0.0000 g, the weight of the container shall be deducted.

(4) Turn off: When weighing is completed, remove the things to be measured, click OFF button to turn off the balance.

3. Apparatus and Materials

(1) Apparatus   electronic balance, a weighting bottle, a beaker (50 mL or 100 mL) or a conical flask (250 mL), stainless steel medicine spoon.

(2) Materials $K_2Cr_2O_7$

4. Procedures

(1) Direct weighing

① Place a clean receiving vessel on the balance pan. The mass of the empty vessel is called the tare.

② Add the chemical to the vessel and read its mass.

(2) Weighing by difference (suitable for absorbing water, oxidation or reaction with carbon dioxide)

① First weigh a capped bottle containing dry reagent. Then quickly pour some reagent from the weighing bottle into a receiver.

② Cap the weighing bottle and weigh it again. The difference is the mass of reagent delivered from the weighing bottle.

(3) Weighing of fixed weight (suitable for the stable sample in air)

① Weigh the weighing paper (or weighing bottle, small beaker and so on) alone on the balance.

② Deducting the tare: Press the TAR key, the display appears 0.0000 g, and the container weight is deducted.

③ Add slowly sample with a clean spatula to the container until it meets the weighing requirements.

Weighing by difference

No.	First	Second	Third
Weighting bottle + sample/g			
Sample/g			

5. Notes

(1) To protect the balance from corrosion, chemicals should never be placed directly on the weighing pan.

(2) Weighing range is generally said to take as the required sample weight of $W(1\pm10\%)$.

(3) Note the sample's weight should not exceed the maximum capacity of the balance, in order to avoid the damage to balance.

(4) In addition to the ON/OFF key and TAR key, students are not allowed to touch the rest of the balance keys in this experiment.

6. Questions

(1) In weighing by difference, does the zero point need to be set? Why?

(2) In the process of weighing by difference, can add sample with medicine spoon?

实验 1　电子天平与称量练习

1. 实验目的
（1）学会正确使用电子分析天平。
（2）熟悉加重称量和减重称量的方法。

2. 实验原理
电子天平使用电磁力补偿来平衡底盘上的负载。一般具有自动校准、零点调整、自动皮重、过载指示、故障报警、自动称重等功能。电子天平具有寿命长、性能稳定、操作方便、灵敏度高的优点。以下是电子天平使用步骤的简要说明，以电子天平的常见赛多利斯 BS-224S 型为例。

赛多利斯 BS-224S 型电子天平

（1）开启　点击 ON 按钮，经过简单的自我测试，显示屏幕应该显示"0.0000g"。如果显示屏幕不是"0.0000g"，按一下 STAR 键。

（2）称重　把物品轻轻地放在称重盘上。当显示屏幕上的数字处于稳定状态时，您可以读取并记录称重结果。

（3）扣除皮重　当一个容器放在称重盘上时，它的重量会显示出来。按下 STAR 键，显示出现 0.0000g，容器的重量已扣除。

（4）关闭　称重完毕，拿走已测量的物品，按 OFF 按钮关闭平衡。

3. 仪器和材料
（1）仪器　电子分析天平，称量瓶，烧杯（50mL 或 100mL）或锥形瓶（250mL），不锈钢药品匙。

（2）材料　$K_2Cr_2O_7$。

差减法称量固体

4. 实验步骤
（1）直接称重
① 在称重盘上放置一个清洁的容器。空容器的质量称为皮重。
② 把化学药品加入到容器中，并读取数字，记下药品质量。

（2）减量称重（适用于易吸水、易氧化或易与二氧化碳反应的物质）
① 取一定量的粉末样品于称量瓶中，在天平上精密称量，记录称量瓶中样品的重量。
② 将称量瓶中的样品小心地倒入小烧杯中，再称量一次称量瓶的重量。两者之差即所需样品的重量。减重称量步骤记录在下表。

减重法称量

编号	第一次	第二次	第三次
称量瓶＋样品 /g			
样品 /g			

5. 注意事项

（1）为了保护天平不受腐蚀，化学品不应该直接放在称重的平底盘上。

（2）称重范围通常以要求的样品重量为准（1±10)%。

（3）注意样品的重量不应超过天平的最大容量，以避免破坏天平。

（4）除了 ON/OFF 键和 STAR 键，在实验中，学生不允许触摸其余的天平键。

6. 思考

（1）在减重法称量中，是否需要设置零点？为什么？

（2）在减重法称量中，可以用药匙添加样品吗？

 Exercises

2-1 Define briefly each of the following terms:
(a) titrant (b) standard solution
(c) titration end point (d) equivalence point

2-2 Distinguish the terms of titration end point and equivalence point.

2-3 What is the difference between a direct titration and a back titration?

2-4 What requirements should the primary standard meet?

2-5 Describe the preparation of solution with the volumetric flask.

2-6 In titrimetric analysis, which is the main requirements for chemical reaction?
A. The reaction must be completed quantitatively.
B. Color changes in the reaction.
C. The titrant reacted with the measured object by a 1∶1 metrological relationship.
D. The titrant must be a primary standard.

2-7 What is the point where the indicator color changes?
A. Equivalence point B. Titration end point error
C. Titration end point D. Titrimetric Analysis

2-8 What causes the titration error?
A. That the equivalence point is not the same as the end point.
B. That titration reaction is not complete.
C. That the sample is not pure enough.
D. That burette reading is not accurate.

2-9　Which must be used to prepare the standard solution by direct method?
　　A. Primary standard　　　　　　　B. Chemically pure reagent
　　C. Analytical reagent　　　　　　　D. Superior pure reagent

2-10　Which affects the analysis results when pouring the weighed reference material into the wet beaker?
　　A. positive error　　B. negative error　　C. no effect　　D. uncertain

2-11　The burette can be estimated to read ± 0.01 mL. If the relative error of titration is required to be less than 0.1%, how much the volume will be consumed at least?
　　A. 10　　　　B. 20　　　　C. 30　　　　D. 40

2-12　Which is the titer of 0.2000 mol·L^{-1} NaOH solution to H$_2$SO$_4$?
　　A. 0.00049　　B. 0.0049　　C. 0.00098　　D. 0.0098

2-13　How much concentrated hydrochloric acid (12 mol·L^{-1}) will be consumed to prepare 1000 mL hydrochloric acid solution (0.1 mol·L^{-1})?
　　A. 0.84　　B. 8.4　　C. 1.2　　D. 12

2-14　Which can be used to standardize both NaOH solution and KMnO$_4$?
　　A. H$_2$C$_2$O$_4$·2H$_2$O　　B. Na$_2$C$_2$O$_4$　　C. HCl　　D. H$_2$SO$_4$

2-15　When determining the content of CaCO$_3$, a certain amount of HCl standard solution is added to completely react with it, and the excess HCl is titrated with NaOH solution. Which titration method is used?
　　A. Direct titration　　　　　　　　B. Back titration
　　C. Replacement titration　　　　　　D. Indirect titration

2-16　How many grams of solid NaOH should be weighed to prepare 500 mL of 1 mol·L^{-1} NaOH solution?

(20 g)

2-17　How much water should be added to dilute 2000 mL of standard solution with concentration of 0.1024 mol·L^{-1} to 0.1000 mol·L^{-1} exactly?

(48 mL)

2-18　Calculate the molar concentration of HCl standard solution, if $T_{Na_2CO_3/HCl}$ = 0.005300 g·mL^{-1}.

(0.1000 mol·L^{-1})

2-19　The primary standard of potassium biphthalate (KHC$_8$H$_4$O$_4$) was used to standardize 0.1 mol·L^{-1} HCl solution. When 30.00 mL of NaOH solution was consumed, how many the potassium biphthalate should be weighed?

(0.6127 g)

2-20　A primary standard of Na$_2$CO$_3$ sample weighing 0.1580 g was used to standardize the concentration of HCl solution. When 24.80 mL of HCl solution is consumed, calculate the concentration of HCl solution.

(0.1202 mol·L^{-1})

Chapter 3　Acid-base Titration
第 3 章　酸碱滴定法

 Study Guide　学习指南

Acid-base titration is a titrimetric analysis method based on acid-base reaction. In view of the fact that it can be applicable not only for aqueous systems, but for non-aqueous systems as well, acid-base titration becomes one of the widely-used methods in titrimetric analysis. Since the groundwork of acid-base titration lies in acid-base reaction, this chapter will make a brief introduction on the basic theory of acid-base balance in solution, and then puts emphasis on learning the fundamental principles of acid-base titration as well as its applications. The learning objectives to be mastered in this chapter are as follows:to understand Brønsted–Lowry acid-base theory, the distribution of weak electrolytes with different species changes with pH of the solution, proton condition, the connection between the establishment of proton condition and the pH formula for acid-base solutions, to comprehend the principles of indicator discoloration, to master the basic principles of acid-base titration. The competence target of this charpter are as follows:to calculate the pH values of kinds of acid-base solutions by using the simplest formula, to calculate the distribution coefficients of weak electrolytes with different species, and to understand the stepwise titration of polybasic acids with the help of the distribution coefficients, to calculate the pH value of the stoichiometric point, and to select the correct indicator for acid-base titration, to analyze and solve practical problems and issues by applying the criteria for accurate titration of monoacidic base and for stepwise titration of polybasic acids or bases into use, to understand the application of acid-base titration in non-aqueous solutions.

　　酸碱滴定法是以酸碱反应为基础的滴定分析方法。它不仅能用于水溶液体系，也可用于非水溶液体系，故酸碱滴定法是滴定分析中广泛应用的方法之一。由于酸碱滴定法的基础是酸碱反应，因此本章首先对溶液中酸碱平衡的基本理论作一简介，然后着重学习酸碱滴定法的基本原理及其应用。本章需要掌握的学习目标如下：理解酸碱质子理论，理解弱电解质不同型体的分布随溶液 pH 的变化，理解质子条件，了解质子条件的建立与酸碱溶液 pH 计算公式的联系，了解指示剂的变色原理，掌握酸碱滴定法的基本原理。本章主要的能力目标包括：能够应用最简式计算各类酸碱溶液的 pH，能计算弱电解质不同型体的分布分数，利用分布分数理解多元酸的分步滴定，会计算化学计量点的 pH，并能正确选择酸碱滴定的指示剂，能应用一元酸碱准确滴定、多元酸碱分步滴定的判据，分析和解决实际问题，了解非水溶液中酸碱滴定的应用。

Section 1 Theoretical Grounding for Acid-base Balance
第 1 节 酸碱平衡的理论基础

1. Brønsted-Lowry acid-base theory

In 1923, Brønsted (J. N. Brønsted) proposed the acid-base theory on the basis of the acid-base ionization theory. The acid-base theory points out that **acids as proton donors and bases as proton acceptors.** When a certain acid HA loses a proton and then forms acid radical A^-, it naturally has an affinity for the proton, so A^- is a base.

1. 酸碱质子理论

1923 年,布朗斯特(Brønsted)在酸碱电离理论的基础上,提出了酸碱质子理论。酸碱质子理论认为:凡是能给出质子 H^+ 的物质是酸;凡是能接受质子的物质是碱。当某种酸 HA 失去质子后形成酸根 A^-,它自然对质子具有一定的亲和力,故 A^- 是碱。

acid-base theory
酸碱理论

$$HA + H_2O \rightleftharpoons H_3O^+ + A^-$$
(Acid)　　　　(Proton)　(Base)

HCl is an acid (a proton donor) and it increases the concentration of H_3O^+ in water:

HCl 是一种酸(给出质子),在水中,他是这样提供质子的:

$$HCl + H_2O \rightleftharpoons H_3O^+ + Cl^-$$

Due to the transfer of a proton, HA and A^- form a pair of acid-base that can be converted to each other, which is called a **conjugate acid-base pair**. The products of a reaction between an acid and a base are also classified as acids and bases.

由于一个质子的转移,HA 与 A^- 形成一对能互相转化的酸碱,称为共轭酸碱对。

Acetate is a base because it can accept a proton to make acetic acid. Methylammonium ion is an acid because it can donate a proton and become methylamine. Acetic acid and the acetate ion are said to be a conjugate acid-base pair. Methylamine and methylammonium ion are likewise conjugate. **Conjugate acids and bases are related to each other by the gain or loss of one H^+.**

For example,

$$HOAc \rightleftharpoons H^+ + OAc^-$$

$$HCl \rightleftharpoons H^+ + Cl^-$$

$$HCO_3^- \rightleftharpoons H^+ + CO_3^{2-}$$

$$H_2C_2O_4 \rightleftharpoons H^+ + HC_2O_4^-$$

$$H_2PO_4^- \rightleftharpoons H^+ + HPO_4^{2-}$$

$$NH_4^+ \rightleftharpoons H^+ + NH_3$$

乙酸盐是一种碱，因为它可以接受一个质子生成乙酸。甲基铵离子是一种酸，因为它可以提供一个质子转化为甲基胺。乙酸和乙酸根离子就是一对共轭酸碱对。甲胺和甲基铵离子同样是共轭酸碱对。共轭酸和共轭碱就是通过得失一个 H^+ 而相互联系。

例如，

In the aforementioned each conjugate acid-base pair, gain or loss reaction of protons is called half-reaction of acid-base, which is similar to the half-cell reaction in the redox reaction. The acid-base half-reaction cannot be carried out alone in solution. When an acid is a proton donor, there must have a base in the solution that is the proton acceptor.

For example, when acetic acid (HOAc) dissociates in aqueous solution, the solvent water is the base that accepts protons, and two acid-base pairs interact to reach equilibria. The reaction formula is as follows:

上述各共轭酸碱对的质子得失反应，称为酸碱半反应，与氧化还原反应中的半电池反应相类似。酸碱半反应在溶液中是不能单独进行的。当一种酸给出质子时，溶液中必定有一种碱接受质子。

例如，乙酸（HOAc）在水溶液中解离时，溶剂水就是接受质子的碱，两个酸碱对相互作用而达平衡，反应式如下：

Half-reaction 1 $HOAc \rightleftharpoons H^+ + OAc^-$

Half-reaction 2 $H_2O + H^+ \rightleftharpoons H_3O^+$

Overall reaction

$$\underset{Acid_1}{HOAc} + \underset{Base_2}{H_2O} \rightleftharpoons \underset{Acid_2}{H_3O^+} + \underset{Base_1}{OAc^-}$$

(Conj. pairs: $Acid_1$–$Base_1$; $Base_2$–$Acid_2$)

In the same way, the process of a base accepting protons in aqueous solution must involve solvent molecules. For instance, the reaction between NH_3 and water is as follows:

同样地，碱在水溶液中接受质子的过程也必须有溶剂分子参加。如 NH_3 与水的反应如下：

Half-reaction 1 $\quad\quad\quad\quad NH_3 + H^+ \rightleftharpoons NH_4^+$

Half-reaction 2 $\quad\quad\quad\quad H_2O \rightleftharpoons H^+ + OH^-$

Overall reaction

$$NH_3 + H_2O \rightleftharpoons OH^- + NH_4^+$$
$$\text{Base}_2 \quad \text{Acid}_1 \quad\quad \text{Base}_1 \quad \text{Acid}_2$$
(Conj. / Conj.)

In the equilibria reached between the two acid-base pairs mentioned above, H_2O plays different roles. In the latter equilibria, the solvent water acts as an acid.

In accordance with the acid-base theory, acids or bases can be positive ions, negative ions, or neutral molecules. The same substance, under a given condition, can be an acid while it turns into a base under another condition, which mainly depends on how close their affinity for protons is. For example, HCO_3^- behaves as a base in the H_2CO_3-HCO_3^- system while it acts as an acid in the HCO_3^--CO_3^{2-} system. Such kind of substance that can either donate protons as acids or accept protons as bases is considered as *amphoteric substance*. It has been known from the interaction between HOAc and H_2O and the interaction between NH_3 and H_2O that water is also an amphoteric substance, usually called an *amphoteric solvent*. The transfer of protons between water molecules can also occur. The proton transferring reaction between the water molecules is called *autoprotolysis*, and the equilibrium constant, K_w, is called *autoprotolysis* or self-ionization constant of water.

在上述两个酸碱对相互作用而达的平衡中，H_2O 分子起的作用不相同，在后一个平衡中，溶剂水起了酸的作用。

按照酸碱质子理论，酸碱可以是阳离子、阴离子，也可以是中性分子。同一种物质，在某一条件下可能是酸，在另一条件下可能是碱，这主要取决于它们对质子亲和力的相对大小。例如 HCO_3^- 在 H_2CO_3-HCO_3^- 体系中表现为碱，而在 HCO_3^--CO_3^{2-} 体系中却表现为酸。这种既可以给出质子表现为酸，又可以接受质子表现为碱的物质，称为两性物质。由 HOAc 与 H_2O 的相互作用和 NH_3 与 H_2O 的相互作用可知，水也是一种两性物质，通常称之为两性溶剂。水分子之间也可以发生质子的转移作用。这种在溶剂分子之间发生的质子传递作用，称为溶剂的质子自递反应，该反应的平衡常数称为质子自递常数。水的质子自递常数写作 K_w，常称为水的离子积。

$$H_2O + H_2O \rightleftharpoons H_3O^+ + OH^-$$
(Conj. / Conj.)

In aqueous solution, the hydrated proton is indicated by H_3O^+; however, for the sake of convenience, it is usually written as H^+. As a result, the expression of K_w can be abbreviated as:

在水溶液中，水化质子用 H_3O^+ 表示，但为了简便起见，通常写成 H^+，所以 K_w 的表示式可以简写为

$$K_w = [H_3O^+][OH^-] = [H^+][OH^-] \quad ❶ \tag{3.1}$$

$$K_w = 10^{-14} (25℃)$$

An approximate definition of pH is the negative logarithm of the H⁺ concentration:

pH 的近似定义是 H⁺ 浓度的负对数：

$$pH = -\lg[H^+] \tag{3.2}$$

A useful relation between $[H^+]$ and $[OH^-]$ is

$[H^+]$ 和 $[OH^-]$ 之间的关系是：

$$pH + pOH = -\lg(K_w) = 14.00 \quad \text{at } 25℃ \tag{3.3}$$

Equation (3.3) is a fancy way of saying that if pH=3.58, then pOH=14.00−3.58=10.42, or $[OH^-]=10^{-10.42}=3.8\times10^{-11}$ M (M means mol·L⁻¹).

式（3.3）意味着，若 pH=3.58，则 pOH=14.00−3.58=10.42，或$[OH^-]=10^{-10.42}=3.8\times10^{-11}$M（M 的意思是 mol·L⁻¹）。

A solution is acidic if $[H^+] > [OH^-]$. A solution is basic if $[H^+] < [OH^-]$. At 25℃, an acidic solution has a pH below 7 and a basic solution has a pH above 7.

若 $[H^+] > [OH^-]$，则该溶液为酸。若 $[H^+] < [OH^-]$，则该溶液为碱。25℃时，酸溶液 pH＜7，碱溶液 pH＞7。

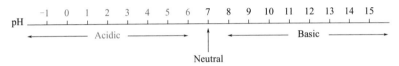

Although pH generally falls in the range 0 to 14, these are not the limits of pH. A pH of −1, for example, means $-\lg[H^+]=-1$, or $[H^+]=10$ M. This concentration is attained in a concentrated solution of a strong acid such as HCl.

虽然 pH 通常在 0～14 范围内，但这不是 pH 值的极值。例如 pH=−1，则 $\lg[H^+]=-1$，$[H^+]=10$ M。浓强酸（如浓 HCl）的 $[H^+]$ 可达到此值。

Example 3-1

Calculate the concentrations of H⁺ and OH⁻ in pure water at 25℃.

Solution

The stoichiometry of reaction tells us that H⁺ and OH⁻ are produced in a 1∶1 molar ratio. Their concentrations must be equal. Calling each concentration x, we can write

$$H_2O \xrightleftharpoons{K_w} H^+ + OH^- \qquad K_w = [H^+][OH^-]$$

$$K_w = [H^+][OH^-] = [x][x] = 1.0\times10^{-14} \Rightarrow x = 1.0\times10^{-7} \text{ M}$$

The concentrations of H⁺ and OH⁻ are both 1.0×10^{-7} M in pure water.

Example 3-2

What is the concentration of OH⁻ if $[H^+]=1.0\times10^{-3}$ M? (Assume that the temperature is 25℃)

---

❶ [ ] means the concentration of certain substance.

> **Solution**
>
> Putting [H$^+$]=1.0×10^{-3} M into the K_w expression gives
>
> $$K_w = 1.0 \times 10^{-14} = (1.0 \times 10^{-3})[OH^-] \Rightarrow [OH^-] = 1.0 \times 10^{-11} M$$
>
> A concentration of [H$^+$]=1.0×10^{-3} M gives [OH$^-$]=1.0×10^{-11} M
> As the concentration of H$^+$ increases, the concentration of OH$^-$ necessarily decreases, and vice versa.
> A concentration of [OH$^-$]=1.0×10^{-3} M gives [H$^+$]=1.0×10^{-11} M.
>
> **TEST YOURSELF**
>
> Find [OH$^-$] if [H$^+$]=1.0×10^{-4} M. (Answer: 1.0×10^{-10} M)

According to the acid-base theory, the essence of acid-base neutralization reaction and salt hydrolysis can also be treated as a process of proton transferring. Thus it can be seen that the acid-base theory reveals the common essence in various acid-base reactions.

根据酸碱质子理论，酸碱中和反应、盐的水解等，其实质也是一种质子的转移过程。可见，酸碱质子理论揭示了各类酸碱反应共同的实质。

2. Dissociation equilibria of acids and bases

2. 酸碱解离平衡

On the basis of the acid-base theory, when an acid or base is added to the solvent, the process of proton transferring will occur and then the corresponding conjugate base or conjugate acid will be generated. The acid dissociation equilibria constant is signified by K_a. For example, when dissociation reaction of HOAc occurs in water:

根据酸碱质子理论，当酸或碱加入溶剂后，就发生质子的转移过程，并生成相应的共轭碱或共轭酸。例如，HOAc 在水中发生解离反应，酸解离平衡常数用 K_a 表示。

$$HOAc + H_2O \rightleftharpoons H_3O^+ + OAc^-$$

$$K_a = \frac{[H^+][OAc^-]}{[HOAc]}, K_a = 1.8 \times 10^{-5}$$

The conjugate base of HOAc is OAc$^-$ and its dissociation constant is K_b:

HOAc 的共轭碱为 OAc$^-$，其解离平衡常数为 K_b：

$$OAc^- + H_2O \rightleftharpoons HOAc + OH^-$$

$$K_b = \frac{[HOAc][OH^-]}{[OAc^-]}$$

Obviously, relationship between K_a and K_b for the acid-base conjugate pair can be expressed as:

显然，共轭酸碱对的 K_a、K_b 关系可以表示为：

$$K_a \times K_b = \frac{[H^+][OAc^-]}{[HOAc]} \times \frac{[HOAc][OH^-]}{[OAc^-]} = [H^+][OH^-] = K_w = 10^{-14}$$

Example 3-3

Dissociation of NH_3, $K_b = 1.8 \times 10^{-5}$ is given, calculate the dissociation constant K_a in the conjugate acid NH_4^+ of NH_3.

Solution

$$NH_4^+ + H_2O \rightleftharpoons H_3O^+ + NH_3$$

$$K_a = \frac{K_w}{K_b} = \frac{10^{-14}}{1.8 \times 10^{-5}} = 5.6 \times 10^{-10}$$

 Example 3-4

K_a for acetic acid is 1.75×10^{-5}. Find K_b for acetate ion.

Solution

Acetic acid is a typical weak acid.

$$\underset{\substack{\text{Acetic acid} \\ (HA)}}{CH_3-C(=O)(O-H)} \rightleftharpoons \underset{\substack{\text{Acetate} \\ (A^-)}}{CH_3-C(=O)(O^-)} + H^+ \quad K_a = 1.75 \times 10^{-5}$$

$$K_b = \frac{K_w}{K_a} = \frac{10^{-14}}{1.75 \times 10^{-5}} = 5.7 \times 10^{-10}$$

The strengths of acidity or basicity rest with the abilities of acids and bases to donate protons or accept protons. The stronger the ability of donating protons a substance has, the higher its acidity presents; and vice versa. Similarly, the stronger the ability of accepting protons a substance has, the higher its basicity present and vice versa. The values of K_a and K_b (see appendix) quantitatively describe the strengths of the acidity or basicity.

In a conjugate acid-base pair, if the acid is apt to donate protons, which means its acidity is stronger, then the proton affinity of the conjugate base gets weaker, which means the difficulty of accepting protons and then its basicity is weaker. For example, although $HClO_4$, H_2SO_4, HCl, HNO_3 are all strong acids with a strong ability to donate protons in aqueous solution ($K_a \gg 1$), their corresponding conjugate bases almost have no ability to obtain protons from H_2O and convert them

酸碱的强弱取决于酸碱本身给出质子或接受质子能力的强弱。物质给出质子的能力越强，其酸性就越强；反之就越弱。同样的，物质接受质子的能力越强，其碱性就越强；反之就越弱。酸碱的解离常数 K_a、K_b（见附录一）的大小，可以定量地说明酸或碱的强弱程度。

在共轭酸碱对中，如果酸越易给出质子，酸性越强，则其共轭碱对质子的亲和力越弱，就不容易接受质子，其碱性就越弱。如 $HClO_4$、H_2SO_4、HCl、HNO_3 都是强酸，它们在水溶液中给出质子的能力很强，$K_a \gg 1$，但它们相应的共轭碱几乎没有能力从 H_2O 中取得质子转化为共轭酸，K_b 小到无法测出。这些共轭碱都是极弱的碱。而 NH_4^+、

into conjugate acids; that is to say, K_b is too small to be measured. Thus, these conjugate bases are extremely weak ones. However, K_a of NH_4^+, HS^- are 5.6×10^{-10} and 7.1×10^{-15} respectively, which are called weak acids; while their conjugate base NH_3 is a strong base and S^{2-} very strong base.

For polyprotic acids, they always dissociate in a stepwise manner in the aqueous solution and have multiple conjugate acid-base pairs whose K_a and K_b exist certain corresponding relations. For instance, $H_2C_2O_4$, a diprotic acid, can be dissociated in two steps:

HS^- 的 K_a 分别为 5.6×10^{-10}、7.1×10^{-15}，是弱酸，它们的共轭碱 NH_3、S^{2-} 则是强碱。

对于多元酸，它们在水溶液中是分级解离的，存在多个共轭酸碱对，这些共轭酸碱对的 K_a 和 K_b 之间也有一定的对应关系。例如，二元酸 $H_2C_2O_4$ 分两步解离：

K_{a1}
$$H_2C_2O_4 \rightleftharpoons H^+ + HC_2O_4^-$$
$$K_{a1} = \frac{[H^+][HC_2O_4^-]}{[H_2C_2O_4]}$$

K_{a2}
$$HC_2O_4^- \rightleftharpoons H^+ + C_2O_4^{2-}$$
$$K_{a2} = \frac{[H^+][C_2O_4^{2-}]}{[HC_2O_4^-]}$$

K_{b1}
$$C_2O_4^{2-} + H_2O \rightleftharpoons OH^- + HC_2O_4^-$$
$$K_{b1} = \frac{[OH^-][HC_2O_4^-]}{[C_2O_4^{2-}]}$$

K_{b2}
$$HC_2O_4^- + H_2O \rightleftharpoons OH^- + H_2C_2O_4$$
$$K_{b2} = \frac{[OH^-][H_2C_2O_4]}{[HC_2O_4^-]}$$

Whereby the abovementioned equilibria it can be obtained:

由此，可得到下面的关系式：

$$K_{a1} \cdot K_{b2} = K_{a2} \cdot K_{b1} = [H^+][OH^-] = K_w \tag{3.4a}$$

As for triprotic acids（三元酸）, the relationship as follows likewise can be got:

对于三元酸，类似的，可得 K_a、K_b 的关系式：

$$K_{a1} \cdot K_{b3} = K_{a2} \cdot K_{b2} = K_{a3} \cdot K_{b1} = [H^+][OH^-] = K_w \tag{3.4b}$$

Section 2 Distribution of Acid-base Species in Aqueous Solution
第 2 节 水溶液中酸碱组分不同形态的分布

In equilibrium system of a weak acid (base), a substance may exist with different species. The concentration of each existing species is called the equilibrium concentration, and the sum of each equilibrium concentration is named the total concentration or the analytical concentration. The fraction of one certain species accounting for in the total concentration is called the distribution fraction of the species, which is represented by the symbol δ. The equilibrium concentration of each existing species is determined by the concentration of hydrogen ions in the solution, so the distribution fraction of each species also changes with changes in hydrogen ions' concentration in the solution. The relation curve between the distribution fraction δ and the pH of the solution is called the distribution curve. By learning the distribution curve, it can not only help us go deep in the understanding of acid-base titration, complexometric titration and precipitation reaction, and has a guiding value in the selection and control of reaction conditions. Now let's look at how to calculate the distribution fractions of monoprotic weak acids, diprotic weak acids and triprotic weak acids and make a discussion on their distribution curves.

在弱酸（碱）的平衡体系中，一种物质可能以多种形态存在，各存在形式的浓度称为平衡浓度，各平衡浓度之和称为总浓度或分析浓度，某一存在形态占总浓度的分数，称为该存在形态的分布分数，用符号 δ 表示。各存在形态平衡浓度的大小由溶液氢离子浓度所决定，因此每种形态的分布分数也随着溶液氢离子浓度的变化而变化。分布分数 δ 与溶液 pH 间的关系曲线称为分布曲线。学习分布曲线，可以帮助我们深入理解酸碱滴定、配位滴定、沉淀反应等过程，并且对于反应条件的选择和控制具有指导意义。现分别对一元弱酸、二元弱酸、三元弱酸分布分数的计算及其分布曲线进行讨论。

1. The Distribution of Monoprotic Weak Acids

1. 一元弱酸的分布

Take HOAc as an example. Due to the dissociation equilibrium of HOAc in aqueous solution, it exists in two species: HOAc and OAc$^-$.
Total concertation, or analytical concentration: c_{HOAc}
Equilibrium concentrations of HOAc and OAc$^-$: [HOAc] [OAc$^-$]
The fraction of HOAc: δ_{HOAc}; The fraction of OAc$^-$: δ_{OAc^-}.
Whereby,

以 HOAc 为例。由于 HOAc 在水溶液中的解离平衡，它以 HOAc 和 OAc$^-$ 两种形态存在。设 c_{HOAc} 为 HOAc 的总浓度，亦称分析浓度；[HOAc]、[OAc$^-$] 分别为 HOAc、OAc$^-$ 的平衡浓度。δ_{HOAc}、δ_{OAc^-} 分别为 HOAc、OAc$^-$ 的分布分数。根据定义：

$$c_{HOAc} = [HOAc] + [OAc^-]$$

$$\delta_{HOAc} = \frac{[HOAc]}{c_{HOAc}} = \frac{[HOAc]}{[HOAc]+[OAc^-]}$$

$$= \frac{1}{1+\frac{[OAc^-]}{[HOAc]}}$$

$$= \frac{1}{1+\frac{K_a}{[H^+]}}$$

$$\delta_{HOAc} = \frac{[H^+]}{[H^+]+K_a} \tag{3.5a}$$

In the same way
$$\delta_{OAc^-} = \frac{K_a}{[H^+]+K_a} \tag{3.5b}$$

Obviously, the sum of all fractions of each species is equal to 1; that is,	显然，各存在形态分布分数之和等于1，即

$$\delta_{HOAc} + \delta_{OAc^-} = 1$$

If using pH as the horizontal axis, δ as the vertical axis, we can get the distribution curve of HOAc as shown in figure 3-1，as can be seen from the figure: When pH<pK_a, HOAc is predominant. When pH>pK_a, OAc⁻ is predominant. When pH=pK_a, HOAc and OAc⁻ are half & half, fractions are 0.5 for both species.	如果以pH为横坐标，以δ_{HOAc}、δ_{OAc^-}为纵坐标作图，得到如图3-1所示HOAc的分布曲线图。当pH<pK_a，HOAc为主要存在形态；当pH>pK_a，OAc⁻为主要存在形态；当pH=pK_a，HOAc与OAc⁻各占一半，两种形态的分布分数均为0.5。

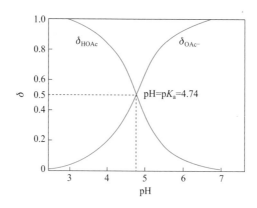

Fig.3-1 HOAc, OAc⁻ distribution fraction curve with solution pH

图3-1 HOAc、OAc⁻分布曲线图

Example 3-5

If the concentration of HOAc is 1.0×10^{-2} mol·L^{-1}. What is the predominant species in the solution when pH=4.0? What is the concentration of it?

Solution

pH=4.0; [H$^+$]=1.0×10^{-4} mol·L^{-1}; K_a=1.8×10^{-5}

Putting [H$^+$]=1.0×10^{-4} mol·L^{-1} and K_a=1.8×10^{-5} into the δ expression gives

$$\delta_{HOAc} = \frac{[H^+]}{[H^+]+K_a} = \frac{1.0 \times 10^{-4}}{1.0 \times 10^{-4} + 1.8 \times 10^{-5}} = 0.85$$

$$\delta_{OAc^-} = \frac{K_a}{[H^+]+K_a} = \frac{1.8 \times 10^{-5}}{1.8 \times 10^{-5} + 1.0 \times 10^{-4}} = 0.15$$

When pH=4.0, HOAc is predominant.

[HOAc]=$c_{HOAc} \times \delta_{HOAc}$=$1.0 \times 10^{-2} \times 0.85$=$8.5 \times 10^{-3}$ mol·L^{-1}

2. The Distribution of Diprotic Weak Acids

2. 二元弱酸的分布

There are three existing species of diprotic weak acids in the solution as shown in figure 3-2: $H_2C_2O_4$, $HC_2O_4^-$ and $C_2O_4^{2-}$. According to the mass balance, the total concentration of oxalic acid should be equal to the sum of the equilibrium concentrations of all species:

如图3-2所示，二元弱酸在溶液中有三种存在形式，即 $H_2C_2O_4$、$HC_2O_4^-$ 和 $C_2O_4^{2-}$。根据质量平衡，草酸的总浓度应等于各形态平衡浓度之和。以 δ 值为纵坐标，以 pH 为横坐标，可得到

$$c_{H_2C_2O_4} = [H_2C_2O_4]+[HC_2O_4^-]+[C_2O_4^{2-}]$$

In terms of definition of fraction:

$$\delta_{H_2C_2O_4} = \frac{[H_2C_2O_4]}{c_{H_2C_2O_4}} = \frac{[H_2C_2O_4]}{[H_2C_2O_4]+[HC_2O_4^-]+[C_2O_4^{2-}]}$$
$$= \frac{1}{1+\frac{[HC_2O_4^-]}{[H_2C_2O_4]}+\frac{[C_2O_4^{2-}]}{[H_2C_2O_4]}} = \frac{1}{1+\frac{K_{a1}}{[H^+]}+\frac{K_{a1}K_{a2}}{[H^+]^2}} \quad (3.6)$$
$$= \frac{[H^+]^2}{[H^+]^2+K_{a1}[H^+]+K_{a1}K_{a2}}$$

In the same way

$$\delta_{HC_2O_4^-} = \frac{K_{a1}[H^+]}{[H^+]^2+K_{a1}[H^+]+K_{a1}K_{a2}} \quad (3.7)$$

$$\delta_{C_2O_4^{2-}} = \frac{K_{a1}K_{a2}}{[H^+]^2+K_{a1}[H^+]+K_{a1}K_{a2}} \quad (3.8)$$

Obviously we can get:

$$\delta_{H_2C_2O_4} + \delta_{HC_2O_4^-} + \delta_{C_2O_4^{2-}} = 1$$

If using pH as the horizontal axis, δ as the vertical axis, we can get the distribution curve of $H_2C_2O_4$ as shown in figure 3-2:

When pH<pK_{a1}, $H_2C_2O_4$ is predominant.
When pH>pK_{a2}, $C_2O_4^{2-}$ is predominant.
When pK_{a1}<pH<pK_{a2}, $HC_2O_4^-$ is predominant.

The distribution curve clearly reflects the relationship between the existing species and the pH of the solution. When selecting the reaction conditions, we can check the Figures on the base of the required components and then can get the corresponding pHs. For example, in order to determine Ca^{2+}, $C_2O_4^{2-}$ used as the precipitant, how much should the pH of the solution be maintained in reaction? Shown in the figure 3-2, when pH $\geqslant$ 5.0, $C_2O_4^{2-}$ is predominant and good for precipitation. So the pH of the solution should be pH $\geqslant$ 5.0.

图 3-2 所示 $H_2C_2O_4$ 的分布曲线图。由图 3-2 可知，当 pH<pK_{a1} 时，$H_2C_2O_4$ 为主要存在形态；当 pH>pK_{a2} 时，$C_2O_4^{2-}$ 为主要存在形态；当 pK_{a1}<pH<pK_{a2} 时，$HC_2O_4^-$ 为主要存在形态。

分布曲线很直观地反映存在形态与溶液 pH 的关系，在选择反应条件时，可以按所需组分查图，即可得到相应的 pH。例如，欲测定 Ca^{2+}，采用 $C_2O_4^{2-}$ 为沉淀剂，反应时，溶液的 pH 应维持在多少？从图 3-2 可知，在 pH $\geqslant$ 5.0 时，$C_2O_4^{2-}$ 为主要存在形态，有利于沉淀形成，所以应使溶液的 pH $\geqslant$ 5.0。

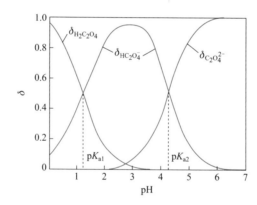

Fig.3-2 The relation curve between the distribution fractions of various existing species in oxalic acid and the pH in the solution

图 3-2 草酸溶液中各种形态的分布分数与溶液 pH 的关系曲线

Example 3-6

When pH=4.0. The concentration of tartaric acid (expressed as H_2A) solution is 5.0×10^{-2} mol·L^{-1}, What is the concentration of tartrate ion A^{2-} in solution?

Solution

pK_{a1}=3.04; pK_{a2}=4.37

Putting $[H^+]=1.0\times 10^{-4}$ mol·L^{-1} and K_{a1}, K_{a2} into the δ expression gives

$$\delta_{A^{2-}} = \frac{K_{a1}K_{a2}}{[H^+]^2 + K_{a1}[H^+] + K_{a1}K_{a2}}$$
$$= \frac{10^{-3.04}\times 10^{-4.37}}{(10^{-4})^2 + 10^{-3.04}\times 10^{-4} + 10^{-3.04}\times 10^{-4.37}}$$
$$= 0.28$$

When pH=4.0, $[A^{2-}]=c_{H_2A}\times \delta_{A^{2-}}=5.0\times 10^{-2}\times 0.28=1.4\times 10^{-2}$ (mol·L^{-1})

3. The distribution of triprotic weak acids

3. 三元弱酸的分布

For triprotic acid like H_3PO_4, it has four existing species: H_3PO_4, $H_2PO_4^-$, HPO_4^{2-}, PO_4^{3-}. In a similar way, we can get the following formula:

三元弱酸如 H_3PO_4 在溶液中有 H_3PO_4、$H_2PO_4^-$、HPO_4^{2-}、PO_4^{3-} 四种形态存在，$\delta_{H_3PO_4}$ 的计算式，见式（3.9）。

$$\delta_{H_3PO_4} = \frac{[H^+]^3}{[H^+]^3 + K_{a1}[H^+]^2 + K_{a1}K_{a2}[H^+] + K_{a1}K_{a2}K_{a3}} \tag{3.9}$$

For the other three species, readers can derive their fraction formulas by taking diprotic weak acids' conditions for reference. The distribution curves of various species in H_3PO_4 solution are shown in the figure 3-3. It should be noted that when pH=4.7, $H_2PO_4^-$ accounts for 99.4%. In the same way, when pH=9.8, HPO_4^{2-} accounts for 99.5% with a compelling advantage.

其余三种形态的分布分数计算式，读者可参照二元弱酸情况自行推出。H_3PO_4 溶液中各种存在形态的分布曲线，如图 3-3 所示。需要指出：在 pH=4.7 时，$H_2PO_4^-$ 形态占 99.4%；同样，当 pH=9.8 时，HPO_4^{2-} 形态占绝对优势，为 99.5%。

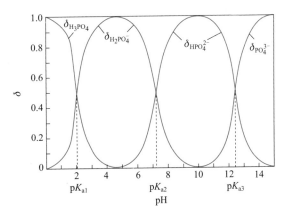

Fig.3-3 The relation curve between the distribution fractions of various existing species in phosphoric acid and the pH in the solution

图 3-3 磷酸溶液中各种形态的分布分数与溶液 pH 的关系曲线

Section 3 pH Calculations for Acid-base Solutions
第 3 节 酸碱溶液 pH 的计算

The process of acid-base titration is the process where the pH of the solution is constantly changing. In order to reveal the changing law of the pH in solution during the titration, this section firstly introduces the calculation methods of pH of several typical acid-base solutions.

酸碱滴定的过程，也就是溶液的 pH 不断变化的过程。为揭示滴定过程中溶液 pH 的变化规律，本节首先学习几类典型酸碱溶液 pH 的计算方法。

1. Calculation of pH of Monoprotic Weak Acid-base Solutions

1. 一元弱酸碱溶液 pH 的计算

Acid-base reaction is the result of proton transfer between substances. The equation is listed according to the quantitative relationship of proton transfer in the overall equilibrium system of acid-base reaction, which is called proton condition, which can calculate the [H$^+$] of the solution. For example, in the aqueous solution of monoprotic weak acid (HA), large amounts of substances present and involved in proton transfer are HA and H$_2$O. The proton transfer reactions in the overall equilibrium system include:

酸碱反应是物质间质子转移的结果。根据酸碱反应整体平衡体系中质子转移的数量关系列出等式，称为质子条件，其可以计算溶液的 [H$^+$]。例如，在一元弱酸（HA）的水溶液中，大量存在并参与质子转移的物质是 HA 和 H$_2$O，整体平衡体系中的质子转移反应有：

Dissociation reaction $\quad$ HA + H$_2$O $\rightleftharpoons$ H$_3$O$^+$ + A$^-$
Autoprotolysis $\quad$ H$_2$O + H$_2$O $\rightleftharpoons$ H$_3$O$^+$ + OH$^-$

Proton conditions can be written based on the amount of matter for the gain and loss of protons that should be equal:

HA 的解离反应和水的质子自递反应，根据得失质子的物质的量应该相等，则质子条件为：

$$[H^+]=[A^-]+[OH^-]$$

Putting $[A^-]=\dfrac{K_a[HA]}{[H^+]}$, $[OH^-]=\dfrac{K_w}{[H^+]}$ into the above formula,
we can get:

将 [A$^-$] 和 [OH$^-$] 的表达式代入质子条件，可得：

$$[H^+]=\dfrac{K_a[HA]}{[H^+]}+\dfrac{K_w}{[H^+]}$$

After sorting, we can get:

简写后，可得：

$$[H^+]=\sqrt{K_a[HA]+K_w} \tag{3.10}$$

The formula (3.10) is the precise formula for calculating [H⁺] in the monoprotic weak acid solution. In the formula, [HA] is the equilibrium concentration of HA, and can be obtained by the formula of distribution fraction, which is quite troublesome. If the calculation of [H⁺] allows an error of 5%, and at the same time, the two conditions of $\frac{c}{K_a} \geq 100, cK_a \geq 10K_w$ (c represents the concentration of monoprotic weak acid) are met, the formula (3.10) can be further simplified as:

$$[H^+] = \sqrt{cK_a} \tag{3.11}$$

Therefore, it becomes the most simplified formula which is commonly used to calculate monoprotic weak acid [H⁺].

式（3.10）为精确计算一元弱酸溶液中 [H⁺] 的公式。式中的 [HA] 为 HA 的平衡浓度，需利用分布分数的公式求得，是相当麻烦的。若计算 [H⁺] 允许有 5% 的误差，同时满足 $\frac{c}{K_a} \geq 100, cK_a \geq 10K_w$（$c$ 表示一元弱酸的浓度）两个条件，式（3.10）可进一步简化为式（3.11）。

这就是计算一元弱酸 [H⁺] 常用的最简式。

 Example 3-7

Calculate pH of 0.10 mol·L⁻¹ HOAc solution.

Solution

Known that $pK_a = 4.74$ and $c = 0.10$ mol·L⁻¹,

$$\frac{c}{K_a} > 100, cK_a > 10K_w$$

[H⁺] can be calculated by the simplified formula.

$$[H^+] = \sqrt{cK_a} = \sqrt{0.10 \times 10^{-4.74}} = 10^{-2.87}$$

As a result, we can get: pH = 2.87

As for monoprotic base solution, by following suit, the most simplified formula of calculating its pH can also be got as long as K_a in the above formula (3.11) is replaced by K_b and [H⁺] by [OH⁻].

对于一元弱碱溶液，同样可以得到计算其 pH 的最简式，即只需将上述计算一元弱酸溶液 [H⁺] 公式（3.11）中的 K_a 换成 K_b，[H⁺] 换成 [OH⁻] 就可以计算弱碱溶液中的 [OH⁻]。

$$[OH^-] = \sqrt{cK_b} \quad (\frac{c}{K_b} \geq 100, cK_b \geq 10K_w) \tag{3.12}$$

Example 3-8

Known that at 25℃, K_a(HOAc)=1.8×10^{-5}. Calculate the concentration of OAc⁻, H⁺ and pH of 0.10 mol·L⁻¹ HOAc solution.

Solution

As a weak electrolyte, the dissociation equilibrium formula of HOAc:

$$HOAc \rightleftharpoons H^+ + OAc^-$$

Initial concentration   0.1   0   0

Equilibrium concentration   0.10−x   x   x

$$K_a = \frac{[H^+][OAc^-]}{HOAc} = \frac{x \cdot x}{0.10-x} = 1.8\times10^{-5}$$

K_a(HOAc) is very low, which can be approximately seen as $0.10-x \approx 0.10$, then we can get:

$$x = \sqrt{1.8\times10^{-5}\times0.10} = \sqrt{1.8\times10^{-6}} = 1.34\times10^{-3} \ (mol \cdot L^{-1})$$

[H⁺]=[OAc⁻]=1.34×10^{-3} mol·L⁻¹

pH = $-\lg[H^+] = -\lg(1.34\times10^{-3}) = 2.87$

Example 3-9

Calculate pH of 0.10 mol·L⁻¹ NH₃·H₂O solution (K_b=1.8×10^{-5}).

Solution

Known that K_b=1.8×10^{-5} and c=0.10 mol·L⁻¹,

$$\frac{c}{K_b} > 100, cK_b > 10K_w$$

[OH⁻] can be calculated by the simplified formula.

$$[OH^-] = \sqrt{cK_b} = \sqrt{0.10\times1.8\times10^{-5}} = 1.3\times10^{-3}$$

As a result, we can get: pOH=2.89

pH=14−pOH=14−2.87=11.13

TEST YOURSELF

Calculate pH of 0.10 mol·L⁻¹ NH₃ solution. Discussion: how to calculate pH of NaOAc solution.

2. Solution of amphiprotic substance

There is a type of substances, such as NaHCO₃, NaH₂PO₄ and potassium acid phthalate, which can not only donate protons showing acidity, but also accept protons displaying basicity. In spite of its complicacy

2. 两性物质溶液

有一类物质，如 NaHCO₃、NaH₂PO₄ 和邻苯二甲酸氢钾等，在水溶液中既可给出质子显示酸性，又可接受质子显示碱性，其酸碱平衡是较为复杂的，

in acid-base balance, in the calculation of [H⁺], a certain reasonably simplified process can also be adopted. Take $NaHCO_3$ as an example, its proton condition is:

$$[H^+]+[H_2CO_3]=[CO_3^{2-}]+[OH^-]$$

K_{a1} and K_{a2}, the equilibrium constant, is taken into the above formula and then after sorting out, we can get:

$$[H^+] = \sqrt{\frac{K_{a1}(K_{a2}[HCO_3^-]+K_w)}{K_{a1}+[HCO_3^-]}} \quad (3.13)$$

If $\frac{c}{K_{a1}} \geq 10$, $cK_{a2} \geq 10K_w$, the formula (3.13) can be simplified as:

$$[H^+] = \sqrt{K_{a1}K_{a2}} \quad (3.14)$$

The formula (3.14) is the most simplified one in calculating pH of amphiprotic substance like NaHA. When the abovementioned two conditions are met, by comparing [H⁺] obtained by the simplified formula with [H⁺] got by the exact equation ($cK_{a2} \geq 10K_w$, $\frac{c}{K_{a1}} \geq 10$), we find relative error falls within the range of 5% which is allowable.

Example 3-10

Calculate pH of $0.10 \text{ mol} \cdot L^{-1}$ NaH_2PO_4 solution.

Solution

Known that pK_{a1}=2.12, pK_{a2}=7.20, pK_{a3}=12.36, and c=0.10 mol·L^{-1},

$\frac{c}{K_{a1}} \geq 10$, $cK_{a2} \geq 10K_w$

[H⁺] can be calculated by the simplified formula.

$$[H^+] = \sqrt{K_{a1}K_{a2}} = \sqrt{10^{-2.12} \times 10^{-7.20}} = 10^{-4.66} \text{ (mol·}L^{-1})$$

As a result, we can get: pH=4.66

Supposing that [H⁺] of Na_2HPO_4 wants to be calculated, K_{a1} and K_{a2} in the equation (3.14) should be replaced by K_{a2} and K_{a3} respectively. Because pH of monoprotic weak acid and amphiprotic substance solutions is frequently required, we put the most simplified formulas as well as their working conditions in the table 3-1 list for calculating pH of various acid solutions.

种酸溶液 pH 的最简式及使用条件列于表 3-1 中。

Table 3-1 Simplified formulas and conditions
表 3-1 计算几种酸溶液 [H^+] 的最简式及使用条件

The type of acid 酸的类型	Simplified formulas 最简式	Conditions 使用条件
Strong acid 强酸	$[H^+] = c$ $[H^+] = \sqrt{K_w}$	$c \geqslant 4.7 \times 10^{-7}$ mol·L^{-1} $c \leqslant 1.0 \times 10^{-8}$ mol·L^{-1}
Monoprotic weak acid 一元弱酸	$[H^+] = \sqrt{cK_a}$	$\dfrac{c}{K_a} \geqslant 100, cK_a \geqslant 10K_w$
Diprotic weak acid 二元弱酸	$[H^+] = \sqrt{cK_{a1}}$	$\dfrac{c}{K_{a1}} \geqslant 100, cK_a \geqslant 10K_w$ $\dfrac{2K_{a2}}{[H^+]} \leqslant 1$
Amphiprotic substance 两性物质	$[H^+] = \sqrt{K_{a1}K_{a2}}$	$\dfrac{c}{K_{a1}} \geqslant 10, cK_{a2} \geqslant 10K_w$

Section 4 Buffer Solution

第 4 节 缓冲溶液

A buffer solution is a solution that changes pH only slightly when small amounts of a strong acid or a strong base are add, or when the solution is slightly diluted. Many buffer solutions are used in analytical chemistry, most of which are used to control the acidity of the solution, and some of which are used as standards for reference when the pH of other solutions is measured, known as buffer standard solutions (table 3-2).

The buffer solution generally consists of a weak acid (or weak base) with a high concentration and its conjugate base (or conjugate acid), such as, HOAc-OAc^-, NH_4^+-NH_3. Since the values of K_{a1} and K_{a2} of the conjugate acid and base are different, the pH ranges that the buffer solution can adjust and control also vary accordingly. Table 3-2 shows the pH ranges that frequently-used buffer solutions display.

能够抵抗外加少量强酸强碱或稍加稀释，其自身 pH 不发生显著变化的性质，称为缓冲作用。具有缓冲作用的溶液称为缓冲溶液。分析化学中要用到很多缓冲溶液，大多数是控制溶液酸度用的，有些则是测量其他溶液 pH 时作为参照标准用的，称为标准缓冲溶液（见表 3-2）。

The use of buffer solution
缓冲溶液的作用

缓冲溶液一般由浓度较大的弱酸（或弱碱）及其共轭碱（或共轭酸）组成。如 HOAc-OAc^-、NH_4^+-NH_3 等。由

Table 3-2 Frequently-used buffer solutions
表 3-2 常用的缓冲溶液

Number 编号	Name 缓冲溶液名称	Existing forms of acids 酸的存在形态	Existing forms of bases 碱的存在形态	pK_a	pH ranges 可控制的 pH 范围
1	Glycine-HCl 氨基乙酸-HCl	$^+NH_3CH_2COOH$	$^+NH_3CH_2COO^-$	2.35 (pK_{a1})	1.4～3.4
2	Monochloroacetic acid-NaOH 一氯乙酸-NaOH	$CH_2ClCOOH$	CH_2ClCOO^-	2.86	1.9～3.9
3	Potassium iphthalate-HCl 邻苯二甲酸氢钾-HCl	(邻苯二甲酸结构，两个COOH)	(邻苯二甲酸氢根结构，COO⁻和COOH)	2.95 (pK_{a1})	2.0～4.0
4	Formic acid-NaOH 甲酸-NaOH	$HCOOH$	$HCOO^-$	3.76	2.8～4.8
5	HOAc-NaOAc	HOAc	OAc^-	4.74	3.8～5.8
6	Hexamethylenetetramine-HCl 六亚甲基四胺-HCl	$(CH_2)_6N_4H^+$	$(CH_2)_6N_4$	5.15	4.2～6.2
7	NaH_2PO_4-Na_2HPO_4	$H_2PO_4^-$	HPO_4^{2-}	7.20 (pK_{a2})	6.2～8.2
8	$Na_2B_4O_7$-HCl	H_3BO_4	$H_2BO_3^-$	9.24	8.0～9.0
9	NH_4Cl-NH_3	NH_4^+	NH_3	9.26	8.3～10.3
10	Glycine-NaOH 氨基乙酸-NaOH	$^+NH_3CH_2COO^-$	$NH_2CH_2COO^-$	9.60	8.6～10.6
11	$NaHCO_3$-Na_2CO_3	HCO_3^-	CO_3^{2-}	10.25	9.3～11.3
12	Na_2HPO_4-NaOH	HPO_4^{2-}	PO_4^{3-}	12.32	11.3～12.0

A buffer solution is composed of weak acid HA and its conjugate base A^-. If c_{HA} and c_{A^-} are used to denote the analytical concentrations of HA and A^- respectively, then $[H^+]$ and pH values in this buffer solution can be calculated in the most simplified equation as follows:

于共轭酸碱对的 K_{a1}、K_{a2} 值不同，所形成的缓冲溶液能调节和控制的 pH 范围也不同，常用的缓冲溶液可控制的 pH 范围参阅表 3-2。

对于由弱酸 HA 与其共轭碱 A^- 组成的缓冲溶液，若用 c_{HA}、c_{A^-} 分别表示 HA、A^- 的分析浓度，可推出计算此缓冲溶液中 $[H^+]$ 及 pH 的最简式，即

$$[H^+] = K_a \cdot \frac{c_{HA}}{c_{A^-}}$$

$$pH = pK_a + \lg\frac{c_{A^-}}{c_{HA}} \qquad (3.15)$$

> **Example 3-11**
>
> Calculate pH of 0.10 mol·L^{-1} HOAc in some certain buffer solution.
>
> Solution
>
> According to equation (3.15), pH can be worked out as follows:
>
> $$\text{pH} = pK_a + \lg\frac{c_{A^-}}{c_{HA}} = -\lg(1.8\times10^{-5}) + \lg\frac{0.15}{0.10} = 4.92$$
>
> TEST YOURSELF
>
> To prepare 1L of buffer solution at pH 10.0, it is known that the solubility of NH$_4$Cl solution is 1.0 mol·L^{-1}. How many milliliters of ammonia water with a density of 0.88 g·mL^{-1} are needed (w_{NH_3}=28%)?   Answer: 380mL

In a strong acid or strong base solution at a high concentration, because the concentration of [H$^+$] or [OH$^-$] is inherently high, the added small amount of acid or base will not have much impact on the acidity of the solution. In this case, strong acid and strong base are also buffer solutions mainly for high acidity (pH<2) and high basicity (pH>12).

Various buffer solutions have different buffering abilities which can be measured by **buffer capacity(β)**. Buffer capacity is the amount of strong acid or base per liter needed to produce a unit change in pH. The larger the buffer capacity of the buffer solution has, the stronger the buffering ability is. Buffer capacity is not only related to the concentration of the components that produce buffering; that is to say, high concentration means great buffer capacity, but also related to the concentration ratio of each component in the buffer solution. In other words, if the total concentration of the buffer components remains constant and the concentration ratio of the buffer components is 1∶1, the buffer capacity reaches the peak. In practical use, the component concentration ratio of weak acids and their conjugate bases is often set as the buffering ranges of the pH (c_a∶c_b=10∶1 and c_a∶c_b=1∶10). From the calculation, we can see:

$$\text{When } c_a:c_b=10:1, \text{pH}=pK_a-1$$
$$\text{When } c_a:c_b=1:10, \text{pH}=pK_a+1$$

Therefore, the buffer range of the buffer solution's pH is

在高浓度的强酸强碱溶液中，由于[H$^+$]或[OH$^-$]的浓度本来就很高，外加的少量酸或碱不会对溶液的酸度产生太大的影响。在这种情况下，强酸强碱也就是缓冲溶液。它们主要是高酸度（pH<2）和高碱度（pH>12）时的缓冲溶液。

各种缓冲溶液具有不同的缓冲能力，其大小可用缓冲容量来衡量。缓冲容量是使1L缓冲溶液的pH增加1个单位所需要加入强碱物质的量，或使溶液pH减少1个单位所需要加入强酸物质的量。缓冲溶液的缓冲容量越大，其缓冲能力越强。缓冲容量的大小与产生缓冲作用组分的浓度有关，其浓度越高，缓冲容量越大。此外，也与缓冲溶液中各组分浓度的比值有关，如果缓冲组分的总浓度一定，缓冲组分的浓度比值为1∶1时，缓冲容量为最大。在实际应用中，常采用弱酸及其共轭碱的组分浓度比分别为c_a∶c_b=10∶1和c_a∶c_b=1∶10，即pH=pK_a-1和pH=pK_a+1作为缓冲溶液pH的缓冲范围。

因而，缓冲溶液pH的缓冲范围为

pH=pK_a±1. For instance, HOAc- and NaOAc solutions show pH=4.74±1, that is to say, the buffer ranges of HOAc and NaOAc change between 3.74 and 5.74 (pH=3.74～5.74). For another example, NH_4Cl-NH_3 can play a buffering role in the range of pH=8.26～10.26.

The pH of the buffer standard solutions is measured through accurate experiments at a certain temperature. Today four buffer standard solutions are specified in the international arena, such as, Saturated Potassium bitartrate at 25℃, 0.05 mol·L^{-1} Potassium biphthalate, 0.025 mol·L^{-1} Potassium dihydrogen phosphate and 0.025 mol·L^{-1} disodium hydrogen phosphate, 0.01 mol·L^{-1} borax. When the acidity conditions need to be strictly controlled in some analyses, buffer standard solutions are applied for monitoring.

Since frequently-used buffer solutions are quite a few, the selection should be done on the ground of actual situations. Cautions: ① The selected buffer solution should not interfere with the analysis process; ② The pH subject to control should be within the buffering ranges of the buffer solution; ③ The concentration ratio of the buffer components should also be 0.01-1 mol·L^{-1} to ensure sufficient buffer capacity. For the preparation of the buffer solution, please consult relevant handbooks or reference books.

pH=pK_a±1。例如，HOAc-NaOAc 缓冲范围为 pH=4.74±1。即 pH=3.74～5.74 为 HOAc-NaOAc 溶液的缓冲范围。又如，NH_4Cl-NH_3 可在 pH=8.26～10.26 范围内起到缓冲作用。

标准缓冲溶液的 pH 是在一定温度下经过准确的实验测得的。目前，国际上规定的标准缓冲溶液有四种（25℃的饱和酒石酸氢钾，0.05mol·L^{-1} 邻苯二甲酸氢钾，0.025mol·L^{-1} 磷酸二氢钾和 0.025mol·L^{-1} 磷酸氢二钠，0.01mol·L^{-1} 硼砂），在某些分析中要严格控制酸度条件时，需要用标准缓冲溶液来监测。

常用缓冲溶液种类很多，要根据实际情况，选用不同的缓冲溶液。注意所选用的缓冲溶液应对分析过程没有干扰，所需控制的 pH 应在缓冲溶液的缓冲范围之内，缓冲组分的浓度也应为 0.01～1mol·L^{-1}，以保证足够的缓冲容量。缓冲溶液的配制，可查阅有关手册或参考书上的配方进行配制。

Section 5 Acid-base Indicators

第5节 酸碱指示剂

An acid-base indicator is itself an acid or base whose various protonated species have different colors. An example is thymol blue. Generally, the color change of an indicator involves establishment of an equilibrium between an acid form and a base form that have different colors.

酸碱指示剂本身是一种酸或碱，其质子化形态不同颜色不同，例如百里酚蓝。即指示剂的颜色会随其酸碱平衡的变化而变化。

Red(R) Thymol blue ⇌ (pK_1=1.7) Yellow(Y$^-$) ⇌ (pK_2=8.9) Blue(B^{2-})

1. Discoloration Range of Indicators

In order to further illustrate the relationship between the color changes of indicators and the acidity, HIn is now used to show the shade of acid color and In$^-$ the shade of base color. The dissociation equilibria of the indicator in the solution is represented by the following formula:

$$HIn \rightleftharpoons H^+ + In^-$$

$$K_{HIn} = \frac{[H^+][In^-]}{[HIn]} \quad (3.16)$$

$$\frac{K_{HIn}}{[H^+]} = \frac{[In^-]}{[HIn]}$$

When $[H^+] = K_{HIn}$, $\frac{K_{HIn}}{[H^+]} = \frac{[In^-]}{[HIn]} = 1$, both of them have the same concentration and the solution presents the intermediate color between acid colors and base colors. At such time, pH=pK_{HIn}, which is called the **theoretical color change point**.

Generally speaking, if $\frac{[In^-]}{[HIn]} > \frac{10}{1}$, the color of In$^-$ can be observed; When $\frac{[In^-]}{[HIn]} = \frac{10}{1}$, the color of HIn can be barely seen in the color of In$^-$, at such time, pH=pK_{HIn}+1.

If $\frac{[In^-]}{[HIn]} < \frac{1}{10}$, the color of HIn can be observed; When $\frac{[In^-]}{[HIn]} = \frac{1}{10}$, the color of In$^-$ can be barely seen in the color of HIn, at such time, pH=pK_{HIn}−1.

From the above discussion, it can be seen that the theoretical discoloration range of indicators is pH= pK_{HIn}±1, which means 2 pH units; however, the actual observed change range of most indicators is less than 2 pH units, and the theoretical discoloration point of the indicator does not fall onto the equivalence point of the discoloration ranges. This is due to the differences in people's sensitivity to different colors. The temperature of the solution also has an effect on the pH transition range of indicators.

Table 3-3 shows commonly used acid-base indicators.

1. 指示剂的变色范围

为了进一步说明指示剂颜色变化与酸度的关系，现以 HIn 表示指示剂酸式形态，以 In$^-$ 代表指示剂碱式形态，在溶液中指示剂的解离平衡常数见式（3.16）。当 [H$^+$]=K_{HIn}，$\frac{K_{HIn}}{[H^+]} = \frac{[In^-]}{[HIn]} = 1$，两者浓度相等，溶液表现出酸式色和碱式色的中间颜色，此时 pH=pK_{HIn}，称为指示剂的理论变色点。一般说来，如果 $\frac{[In^-]}{[HIn]} > \frac{10}{1}$，观察到的是 In$^-$ 的颜色；当 $\frac{[In^-]}{[HIn]} = \frac{10}{1}$ 时，可在 In$^-$ 颜色中勉强看出 HIn 的颜色，此时 pH=pK_{HIn}+1；当 $\frac{[In^-]}{[HIn]} < \frac{1}{10}$ 时，观察到的是 HIn 的颜色；当 $\frac{[In^-]}{[HIn]} = \frac{1}{10}$ 时，可在 HIn 颜色中勉强看出 In$^-$ 的颜色，此时 pH=pK_{HIn}−1。

由上述讨论可知，指示剂的理论变色范围为 pH=pK_{HIn}±1，为 2 个 pH 单位，但实际观察到的大多数指示剂的变化范围小于 2 个 pH 单位，且指示剂理论变色点不是变色范围的中间点，这是由于人们对不同颜色的敏感程度的差别造成的，溶液的温度也影响指示剂的变色范围。

常用的酸碱指示剂见表 3-3。

acid-base indicators
指示剂的配制方法

Table 3-3 Commonly used acid-base indicators
表3-3 常用的酸碱指示剂

Indicator 指示剂	Acid color 酸式色	Base color 碱式色	pK_a	pH transition range	Usage 用法
Thymol blue (discoloration for the first time) 百里酚蓝（第一次变色）	Red	Yellow	1.6	1.2-2.8	0.1% in 20% ethanol 0.1% 的 20% 乙醇
Methyl yellow 甲基黄	Red	Yellow	3.3	2.9-4.0	0.1% in 20% ethanol 0.1% 的 20% 乙醇
Methyl orange 甲基橙	Red	Yellow	3.4	3.1-4.4	0.05% aqueous solution 0.05% 的水溶液
Bromphenol blue 溴酚蓝	Yellow	Violet	4.1	3.1-4.6	0.1% in 20% ethanol or sodium salt solution 0.1% 的 20% 乙醇或其钠盐
Bromcresol green 溴甲酚绿	Yellow	Blue	4.9	3.8-5.4	0.1% aqueous solution, add 0.05 mol·L^{-1} NaOH 9mL per 100mg of indicator. 0.1% 水溶液，每100mg 指示剂加 0.05mol·L^{-1} NaOH 9mL
Methyl red 甲基红	Red	Yellow	5.2	4.4-6.2	0.1% in 60% ethanol or sodium salt solution 0.1% 的 60% 乙醇或其钠盐水溶液
Bromthymol blue 溴百里酚蓝	Yellow	Blue	7.3	6.0-7.6	0.1% in 20% ethanol or sodium salt solution 0.1% 的 20% 乙醇或其钠盐水溶液
Neutral red 中性红	Red	Yellowish orange	7.4	6.8-8.0	0.1% in 60% ethanol 0.1% 的 60% 乙醇
Phenol red 酚红	Yellow	Red	8.0	6.7-8.4	0.1% in 60% ethanol or sodium salt solution 0.1% 的 60% 乙醇或其钠盐水溶液
Thymol blue (discoloration for the second time) 百里酚蓝（第二次变色）	Yellow	Blue	8.9	8.0-9.6	0.1% in 20% ethanol 0.1% 的 20% 乙醇
Phenolphthalein 酚酞	Colorless	Red	9.1	8.0-9.6	0.1% in 90% ethanol 0.1% 的 90% 乙醇
Thymolphthalein 百里酚酞	Colorless	Blue	10.0	9.4-10.6	0.1% in 90% ethanol 0.1% 的 90% 乙醇

Section 6 Titrations of Monoprotic Acid-base
第 6 节 一元酸碱的滴定

The curve shown in fig.3-4 is a typical titration curve for the titration of a strong monoprotic acid(such as HCl) with a strong base(such as NaOH). The dashed lines in this graph indicate the volume of titrant that is needed to reach the equivalence point and show the pH at this point in the titration. A titration in which the measured response makes use of a logarithmic expression of concentration or activity（such as pH）and is known as a logarithmic titration curve. A logarithmic titration curve often has the curved behavior that is shown in this example. It is also possible to have a measured response such as absorbance that is directly related to an analyte's concentration or activity; this second type of plot is known as a linear titration curve.

In the process of acid-base titration, as the titrant is continuously added to the solution to be titrated, the pH of the solution changes correspondingly. Based on the

图 3-4 所示的曲线是强碱（如 NaOH）滴定强酸（如 HCl）的典型滴定曲线。图中虚线表示达到化学计量点所需的滴定剂的体积及溶液此时的 pH 值。滴定分析中，响应信号用浓度或活性的对数来表达的（如 pH 值），称为对数滴定曲线，对数滴定曲线通常具有本例所示的曲线性质；若响应信号与分析物的浓度或活性正相关，如吸光度，这种类型的滴定曲线称为线性滴定曲线。

酸碱滴定过程中，随着滴定剂不断地加入到被滴定溶液中，溶液的 pH 不断地变化，根据滴定过程中溶液 pH 的

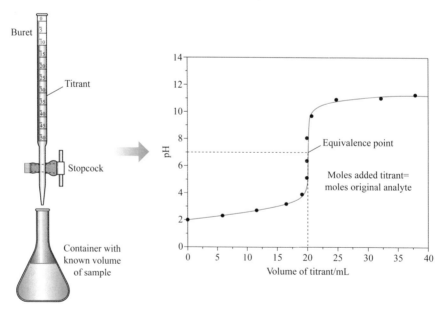

Fig.3-4 A general example of the equipment used in an acid-base titration, and a typical titration curve

图 3-4 酸碱滴定装置及典型的滴定曲线

changing law of the pH of the solution during the titration process, selecting the appropriate indicator can correctly indicate the end point of the titration. This section discusses the law of pH change and the principles of indicator selection in the process of monoprotic acid-base titration.

1. Titration of Strong Base with Strong Acid

Supposing that 20.00mL 0.1000 mol·L^{-1} HCl is titrated with 0.1000 mol·L^{-1} NaOH. The initial pH of ready-to-be titrated HCl solution is low. However, with the addition of NaOH, the neutralization reaction unceasingly takes place with increasing of pH in a row. When the added amount of NaOH is exactly equals to that of HCl, the neutralization reaction just right completes and the titration reaches the stoichiometric point ($[H^+]=[OH^-]=10^{-7}$ mol·L^{-1}). There is only NaCl in the solution. After exceeding the stoichiometric point, with NaOH being added consistently, the pH continues to rise. For a better understanding of the details of the entire titration process, four stages are marked off as follows:

(1) Before the titration　The pH of the solution depends on the original concentration of HCl, that is, the analytical concentration. Since HCl is a strong acid, so $[H^+]=0.1000$ mol·L^{-1}, and pH=1.00.

(2) Before the equivalence point　The pH of the solution is determined by the amount of remaining HCl. For example, if adding 19.98mL of NaOH, then we can get:

$$[H^+] = \frac{c_{HCl} \times V_{HCl}(\text{Volume of Remaining HCl})}{\text{Total Volume of Solution}} = \frac{0.1000 \text{mol} \cdot L^{-1} \times 0.02 \text{mL}}{20.00 \text{mL} + 19.98 \text{mL}}$$

$$= 5.0 \times 10^{-5} \text{ mol} \cdot L^{-1}$$

$$pH = 4.3$$

The pH of other points can follow the same way.

(3) At the equivalence point　At the equivalence point, NaOH and HCl are exactly right neutralized. At such time, $[H^+]=[OH^-]=10^{-7}$ mol·L^{-1}. Therefore, the pH at the equivalence point reaches 7.0 and the solution remains neutral.

(4) After the equivalence point   At this point, the pH of the solution is calculated based on the amount of excess base. If 20.02mL of NaOH is dropped, it means an excess of 0.1%.

（4）化学计量点后　此时溶液的 pH 根据过量碱的量进行计算。如滴入 NaOH 溶液 20.02mL，即过量 0.1%。

$$[OH^-] = \frac{c_{NaOH} \times V_{NaOH}(\text{Volume of Excess NaOH})}{\text{Total Volume of Solution}} = \frac{0.1000 \text{mol} \cdot L^{-1} \times 0.02 \text{mL}}{20.00 \text{mL} + 20.02 \text{mL}}$$
$$= 5.0 \times 10^{-5} \text{mol} \cdot L^{-1}$$
$$pOH = 4.3, pH = 9.7$$

Other points after the equivalence point can also be made in the same way. The above values are listed in table 3-4. With the added volume of NaOH as the abscissa and the corresponding pH as the ordinate, the pH-V relation curve is drawn, which is called the titration curve, as shown in figure 3-5.

化学计量点后的各点，均可按此方法逐一计算。将上述计算值列于表 3-4，以 NaOH 加入量为横坐标、对应的 pH 为纵坐标，绘制 pH-V 关系曲线，称为滴定曲线，如图 3-5 所示。

Table 3-4   20.00mL 0.1000mol·L^{-1} HCl is titrated by 0.1000mol·L^{-1} NaOH
表 3-4　用 0.1000mol·L^{-1} NaOH 溶液滴定 20.00mL 0.1000mol·L^{-1} HCl 溶液

Add NaOH solution 加入 NaOH 溶液		Remaining volume of HCl 剩余 HCl 的体积 V/mL	Excess volume of NaOH 过量的 NaOH 的体积 V/mL	pH
$\alpha^{①}$%	V/mL			
0	0.00	20.00		1.00
90.0	18.00	2.00		2.28
99.0	19.80	0.20		3.30
99.9	19.98	0.02		4.3 (A) ⎫
100.0	20.00	0.00		7.00 ⎬ Titrated jump
100.1	20.02		0.02	9.7 (B) ⎭
101.0	20.20		0.20	10.70
110.0	22.00		2.00	11.70
200.0	40.00		20.00	12.50

① α is the titration fraction and can be defined as:
$$\alpha = \frac{\text{the amount of substance with base being added}}{\text{the amount of substance with initial acidity}}$$

α 为滴定分数，其定义为 $\alpha = \dfrac{\text{加入碱的物质的量}}{\text{酸起始的物质的量}}$

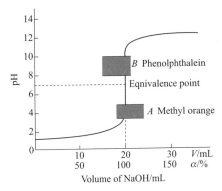

Fig.3-5 The titration curve of 20.00mL 0.1000mol·L⁻¹ HCl is titrated by 0.1000mol·L⁻¹ NaOH

图 3-5　0.1000mol·L⁻¹ NaOH 滴定 20.00mL 0.1000 mol·L⁻¹ HCl 的滴定曲线

It can be seen from table 3-4 and figure 3-5 that the curve is relatively flat at the beginning of the titration which owes to the fact that there is still more HCl in the solution and the acidity is greater. With the continuous dripping of NaOH, the amount of HCl gradually decreases and the pH gradually increases. When the titration leaves only 0.1% HCl, that is, 0.02mL HCl, the pH is 4.3; and then 1 drop of titrant (about 0.04mL) keeps to be added to neutralize the remaining half drop of HCl. After that, only 0.02mL NaOH is excess, and the pH of the solution rises sharply from 4.3 to 9.7. Therefore, one drop of the solution makes the pH of the solution increase by more than 5 pH units. From the point A to B in figure 3-6 and table 3-4, it can be seen that prior to and past the equivalence point 0.1%,

从表 3-4 和图 3-5 可见，滴定开始时曲线比较平坦，这是因为溶液中还存在着较多的 HCl，酸度较大。随着 NaOH 不断滴入，HCl 的量逐渐减少，pH 逐渐增大。当滴定至只剩下 0.1%HCl，即剩余 0.02mL HCl 时，pH 为 4.3。再继续滴入 1 滴滴定剂（大约 0.04mL），即中和剩余的半滴 HCl 后，仅过量 0.02mL NaOH，溶液的 pH 从 4.3 急剧升高到 9.7。因此，1 滴溶液就使溶液 pH 增加 5 个多 pH 单位。从图 3-6 和表 3-4 的 A 点至 B 点可知，在化学计量点前后 0.1%，滴定曲线上出现了一段垂直线，这称为滴定突跃。指示剂的选择主要以滴定突跃为依据，凡在

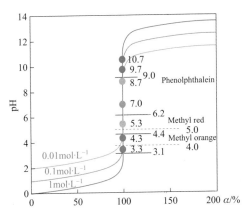

Fig.3-6 Titration curves of different concentrations of HCl solutions titrated by different concentrations of NaOH solutions

图 3-6　不同浓度 NaOH 溶液滴定不同浓度盐酸溶液的滴定曲线

a vertical line appears on the titration curve, which is called a **titration jump**. The choice of indicators is mainly based on the titration jump. Anything that discolors within pH=4.3 ~ 9.7, such as methyl orange, methyl red, phenolphthalein, bromothymol blue, phenol red, etc., can be used as indicators for this type of titration.

For example, when methyl orange changes from red to orange, the pH of the solution is about 4.4. At this time, the difference between the amount of NaOH added and the amount that should be added at the equivalence point is less than 0.02mL, and the end point error is less than −0.1%, which meets the requirements of titration analysis. If phenolphthalein is used as the indicator, the pH of the solution is slightly higher than 8.0 when the solution is reddish. At this moment, the amount of NaOH added exceeds that of at the equivalence point but is still less than 0.02mL and the end point error is also less than +0.1%, which is in line with the titration analysis' requirements. Therefore, **selecting an indicator whose color change range falls into or partly within the titration jump range can accurately indicate the titration end point**, which can be treated as the principle when indicators are selected. This is also an important conclusion in this section.

The above discussion talks about how $0.1000\ mol\cdot L^{-1}$ HCl is titrated by $0.1000\ mol\cdot L^{-1}$ NaOH. If the concentration of the NaOH solution is changed and the pH of the equivalence point remains 7.0, the magnitude of titration jump will be different. As shown in figure 3-6, the higher the concentration of the acid-based solution is, the longer of titration jump is in the vicinity of the equivalence point of the titration curve which means more indicators can be available for selection. On the contrary, the weaker the concentration of the titrant solution is, the shorter the titration jump is in the vicinity of the equivalence point, which indicates the fewer indicators can be chosen from and the choice of indicators is limited. For example, when $0.0100\ mol\cdot L^{-1}$ HCl is titrated by $0.0100\ mol\cdot L^{-1}$ NaOH, the titration jump is reduced to 5.3 ~ 8.7. Supposing that methyl orange is still used as the indicator, the end point error will be larger than

pH=4.3 ～ 9.7 内变色的，如甲基橙、甲基红、酚酞、溴百里酚蓝、苯酚红等，均能作为此类滴定的指示剂。

例如，当滴定至甲基橙由红色突变为橙色时，溶液的 pH 约为 4.4，这时加入 NaOH 的量与化学计量点时应加入量的差值不足 0.02mL，终点误差小于 −0.1%，符合滴定分析的要求。若改用酚酞为指示剂，溶液呈微红色时 pH 略大于 8.0，此时 NaOH 的加入量超过化学计量点时应加入的量也不到 0.02mL，终点误差也小于 +0.1%，仍然符合滴定分析的要求。因此，选择变色范围处于或部分处于滴定突跃范围内的指示剂，都能够准确地指示滴定终点，这是正确选择指示剂的原则，也是本节的一个重要结论。

以上讨论的是 $0.1000mol\cdot L^{-1}$ NaOH 溶液滴定 $0.1000mol\cdot L^{-1}$ HCl 溶液的情况。如改变 NaOH 溶液浓度，化学计量点的 pH 仍然是 7.0，但滴定突跃的长短却不同。由图 3-6 所示，酸碱溶液浓度越大，滴定曲线化学计量点附近的滴定突跃越长，可供选择的指示剂越多。如滴定剂溶液的浓度越小，则化学计量点附近的滴定突跃就越短，可供选择的指示剂就越少，指示剂的选择就受到限制。例如，若用 $0.0100\ mol\cdot L^{-1}$ NaOH 溶液滴定 $0.0100\ mol\cdot L^{-1}$ HCl 溶液，滴定突跃减小为 5.3 ~ 8.7，若仍用甲基橙作指示剂，终点误差将大于 1%，此时只能用酚酞、甲基红等，才能符合滴定分析的要求。

1%, which is to say, only phenolphthalein, methyl red or others can meet the requirements of titration analysis.

The titration of a strong monoprotic acid with a strong base and the titration of a strong monoprotic base with a strong acid are similar. Their equations are shown in fig. 3-7 and fig.3-8.

用强碱滴定强酸的过程与强酸滴定强碱很相似,各阶段计算公式见图 3-7、图 3-8。

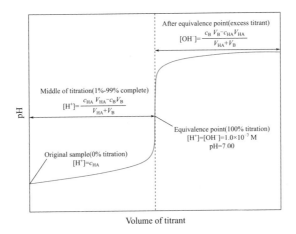

Fig.3-7　Equations for predicting the response for the titration of a strong monoprotic acid with a strong base in water

图 3-7　强碱滴定强酸的反应计算方程式

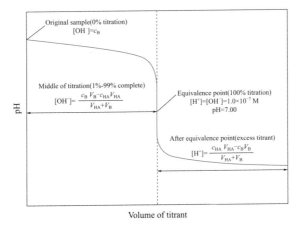

Fig.3-8　Equations for predicting the response for the titration of a strong monoprotic base with a strong acid in water

图 3-8　强酸滴定强碱的反应计算方程式

Example 3-12

When 20.00 mL 0.0200 mol·L^{-1} KOH is titrated by 0.1000 mol·L^{-1} HBr, (1) During the titration, how does the pH of the solution change? (2) How about titration jump? (3) Which indicator can be selected to meet the requirement of titration error which shall be less than $\pm 0.1\%$?

2. Titration of Weak Acid with Strong Base

In the case of weak acid titrated by strong base, for example, if 0.1000 mol·L^{-1} NaOH titrates 20.00 mL 0.1000 mol·L^{-1} HOAc, during the titration, the pH of the solution can be calculated as follows. (pK_a=4.74 of HOAc is known)

(1) Before the titration The pH of the solution is calculated based on the HOAc dissociation equilibrium:

$$[H^+] = \sqrt{cK_a} = \sqrt{0.10 \times 10^{-4.74}} = 10^{-2.87} \qquad pH=2.87$$

(2) Before the equivalence point The pH of the solution at this stage should be calculated according to the formula (3.15) based on the buffer solution composed of the remaining HOAc and the reaction product OAc$^-$. Now it is assumed that 19.98mL of NaOH is added. Then NaOAc is formed after neutralizing with HOAc and 0.02mL of remaining HOAc is left without being neutralized. pH is calculated as follows:

$$[HOAc] = \frac{0.1000 \text{mol} \cdot \text{L}^{-1} \times 0.02 \text{mL}}{20.00 \text{mL} + 19.98 \text{mL}} = 5.0 \times 10^{-5} \text{mol} \cdot \text{L}^{-1}$$

$$[OAc^-] = \frac{0.1000 \text{mol} \cdot \text{L}^{-1} \times 19.98 \text{mL}}{20.00 \text{mL} + 19.98 \text{mL}} = 5.0 \times 10^{-2} \text{mol} \cdot \text{L}^{-1}$$

$$[H^+] = K_a \cdot \frac{c_{HOAc}}{c_{OAc^-}} = 1.8 \times 10^{-5} \times \frac{5.0 \times 10^{-5}}{5.0 \times 10^{-2}} = 1.8 \times 10^{-8} (\text{mol} \cdot \text{L}^{-1}) \qquad pH=7.7$$

(3) At the equivalence point NaOH and HOAc are completely neutralized, and the reaction product is NaOAc. Based on the dissociation equilibrium of the conjugate base, we can calculate as follows:

$$OAc^- + H_2O \longrightarrow HOAc + OH^-$$

$$[OAc^-] = \frac{0.1000 \text{mol} \cdot \text{L}^{-1} \times 20.00 \text{mL}}{20.00 \text{mL} + 20.00 \text{mL}} = 5.000 \times 10^{-2} \text{mol} \cdot \text{L}^{-1}$$

$$[OH^-] = \sqrt{cK_b} = \sqrt{c \times \frac{K_w}{K_a}} = \sqrt{5.000 \times 10^{-2} \times \frac{1.0 \times 10^{-14}}{1.8 \times 10^{-5}}} = 5.3 \times 10^{-6} (\text{mol} \cdot \text{L}^{-1})$$

pOH=5.28,pH=8.72

2. 强碱滴定弱酸

现以 0.1000mol·L^{-1}NaOH 溶液滴定 20.00mL 0.1000mol·L^{-1}HOAc 溶液为例，讨论强碱滴定弱酸的情况，滴定过程中溶液 pH 可计算如下。已知 HOAc 的解离常数 pK_a=4.74。

（1）滴定开始前 溶液 pH 可由下式求得：

（2）滴定至化学计量点前 这阶段溶液的 pH 应根据剩余的 HOAc 及反应产物 OAc$^-$ 所组成的缓冲溶液按式（3.15）计算。现设滴入 NaOH 19.98mL，与 HOAc$^-$ 中和后形成 NaOAc，剩余 HOAc 0.02mL 未被中和。

（3）化学计量点时 NaOH 和 HOAc 完全中和，反应产物是 NaOAc，根据共轭碱的解离平衡，可计算如下：

(4) After the equivalence point　　　　　　（4）化学计量点后

$$[OH^-] = \frac{0.1000 \text{mol} \cdot L^{-1} \times 0.02 \text{mL}}{20.00 \text{mL} + 20.02 \text{mL}} = 5.0 \times 10^{-5} \text{mol} \cdot L^{-1}$$

pOH=4.3，pH=9.7

Equations for predicting the response for the titration of a weak monoprotic acid with a strong base are shown in fig.3-9.

强碱滴定弱酸的反应计算方程式见图 3-9。

The above results are listed in table 3-5. The titration curve drawn according to the values in table 3-5 is as shown in figure 3-10.

以上结果见表 3-5。依据表 3-5 数据，可绘制滴定曲线（图 3-10）。

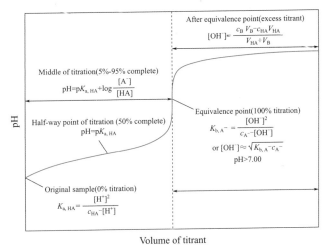

Fig.3-9　Equations for predicting the response for the titration of a weak monoprotic acid with a strong base

图 3-9　强碱滴定弱酸的反应计算方程式

Table 3-5　20.00mL 0.1000 mol·L⁻¹ HOAc is titrated by 0.1000 mol·L⁻¹ NaOH

表 3-5　用 0.1000mol·L⁻¹ NaOH 溶液滴定 20.00mL 0.1000 mol·L⁻¹ HOAc 溶液

Add NaOH		Remaining Volume of HOAc V/mL	Excess Volume of NaOH V/mL	pH
a/%	V/mL			
0	0.00	20.00		2.87
50.0	10.0	10.00		4.74
99.0	18.00	2.00		5.70
99.0	19.80	0.20		6.74
99.9	19.98	0.02		7.70
100.0	20.00	0.00		8.72
100.1	20.02		0.02	9.70
101.0	20.20		0.20	10.70
110.0	22.00		2.00	11.70
200.0	40.00		20.00	12.50

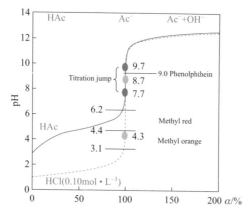

Fig. 3-10 Titration curves of HOAc titrated by NaOH

图 3-10 氢氧化钠滴定 HOAc 的滴定曲线

Comparing the titration curve of HOAc with NaOH and the one of HCl with NaOH, we can find that they have the following differences:

① Because HOAc is a weak acid, before titration, H^+ concentration in the solution is lower than that of HCl with the same concentration, so the initial pH is higher.

② Before the equivalence point, the unreacted HOAc in the solution and the reaction product NaOAc constitutes HOAc-OAc$^-$, so the pH changes relatively slowly.

③ In the vicinity of the equivalence point, the pH of the solution has a sudden change, and the titration jump is pH=7.7～9.7. Compared with the titration of HCl, the magnitude of titration jump is much smaller.

④ At the equivalence point, the solution only contains NaOAc, which is a base substance with a pH of 8.72, so the solution is base at the equivalence point.

Notice:
① When a weak acid is titrated by a strong base, the titration jump range is small, thus the choice of indicator is under restrictions; namely, only indicators that change colors in the range of weak bases can be selected, such as phenolphthalein, thymolphthalein, etc (fig.3-11). If an indicator that changes color in the acid range is still chosen like methyl orange, when the solution changes color, the neutralized fraction of HOAc is less than 50%. Obviously, a wrong indicator is selected. To titrate weak acids, the pH at the equivalence point will be calculated

将 NaOH 滴定 HOAc 的滴定曲线与 NaOH 滴定 HCl 的滴定曲线相比较，可以看到它们有以下不同点。

① 由于 HOAc 是弱酸，滴定前溶液中 H^+ 浓度比同浓度的 HCl 的 H^+ 浓度要低，因此起始的 pH 要高一些。

② 化学计量点之前，溶液中未反应的 HOAc 与反应产物 NaOAc 组成了 HOAc-OAc$^-$，pH 的变化相对较缓。

③ 化学计量点附近，溶液的 pH 发生突变，滴定突跃为 pH=7.7～9.7，相对滴定 HCl 而言，滴定突跃小得多。

④ 化学计量点时，溶液中仅含 NaOAc，为一碱性物质，pH 为 8.72，因而化学计量点时溶液呈碱性。

注意：
① 强碱滴定弱酸时，滴定突跃范围较小，指示剂的选择受到限制，只能选择在弱碱性范围内变色的指示剂，如酚酞、百里酚酞等（图 3-11）。若仍选择在酸性范围内变色的指示剂，如甲基橙，溶液变色时，HOAc 被中和的分数还不到 50%，显然，指示剂选择错误。滴定弱酸，一般都是先计算出化学计量点时的 pH，选择那些变色点尽可能接近化学计量点的指示剂来确定终点，而

firstly, and the indicator whose color changing point is as close as possible to the equivalence point will be selected to determine the end point, instead of calculating the pH changes in the entire titration process.

② The magnitude of the titration jump when the weak acid is titrated by a strong base is up to the concentration of the weak acid solution and its dissociation constant K_a. If the titration error is required to be less than or equal to ±0.1%, the titration jump must be over 0.3 pH units. At this time, the color changes of the indicator become visible and the titration can be proceeded smoothly. It can be seen from figure 3-12 that the weak acid with a concentration of 0.1000 mol·L^{-1} and $K_a=10^{-6}$ can also show a titration jump of 0.3 pH units. As for the weak acid with $K_a=10^{-8}$, if its concentration is 0.1000 mol·L^{-1}, it cannot be directly titrated visually. Generally, $cK_a \geqslant 10^{-8}$ is used as the criterion for the direct and accurate titration of weak acid by strong base.

② 强碱滴定弱酸时的滴定突跃大小，决定于弱酸溶液的浓度和它的解离常数 K_a 两个因素。如要求滴定误差小于等于±0.1%，必须使滴定突跃超过0.3pH单位，此时人眼才可以辨别出指示剂颜色的变化，滴定就可以顺利地进行。由图3-12可以看出，浓度为0.1000mol·L^{-1}，$K_a=10^{-6}$的弱酸还能出现0.3pH单位的滴定突跃。对于$K_a=10^{-8}$的弱酸，其浓度若为0.1000mol·L^{-1}不能目视直接滴定。通常，以$cK_a \geqslant 10^{-8}$作为弱酸能被强碱溶液直接目视准确滴定的判据。

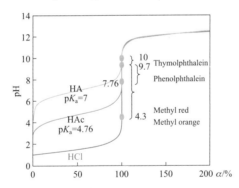

Fig.3-11 Titration curves of different acid solutions titrated by NaOH

图3-11 氢氧化钠滴定不同酸溶液的滴定曲线

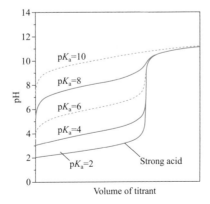

Fig.3-12 Titration curves of different acid solutions titrated by NaOH

图3-12 氢氧化钠滴定不同酸溶液的滴定曲线

3. Titration of Weak Base with Strong Acid

When weak base is titrated by strong acid, for example, the HCl is used to titrate NH_3. The titration reaction is:

$$NH_3 + H^+ \rightleftharpoons NH_4^+$$

With the addition of HCl, the composition of the solution undergoes a change from NH_3 to NH_4^+-NH_3, and then to NH_4Cl with the pH gradually changing from high to low. This type of titration is very similar to that of HOAc with NaOH. We still divide the whole thinking into four stages (equations are shown in fig.3-13) and the specific calculation results are listed in table 3-6, and the titration curve is shown in figure 3-14.

3. 强酸滴定弱碱

强酸滴定弱碱，以 HCl 溶液滴定 NH_3 溶液即属此例，滴定反应为：

随着 HCl 的滴入，溶液组成经历 NH_3 到 NH_4^+-NH_3，再到 NH_4Cl 的变化过程，pH 亦逐渐由高向低变化，这类滴定与用 NaOH 滴定 HOAc 十分相似。现仍采取分四个阶段的思路，相关方程见图 3-13，将具体计算结果列于表 3-6，其滴定曲线如图 3-14 所示。

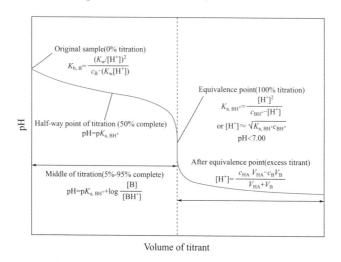

Fig.3-13 Equations for predicting the response for the titration of a weak monoprotic base with a strong acid in water

图 3-13 强酸滴定弱碱的反应计算方程式

Table 3-6 20.00mL 0.1000mol·L^{-1} NH_3 is titrated by 0.1000mol·L^{-1} HCl

表 3-6 用 0.1000mol·L^{-1} HCl 溶液滴定 20.00mL 0.1000mol·L^{-1} NH_3 溶液

Add HCl		Composition	[OH$^-$]or [H$^+$] formulas	pH
V/mL	a/%			
0.00	0	NH_3	$[OH^-] = \sqrt{cK_b}$	11.13
18.00	90.00	NH_4^+-NH_3	$[OH^-] = K_b \cdot \dfrac{c_{NH_3}}{c_{NH_4^+}}$	8.30
19.98	99.9			6.3
20.00	100.0	NH_4^+	$[H^+] = \sqrt{\dfrac{K_w}{K_b} \cdot c_{NH_4^+}}$	5.28
20.02	100.1	NH_4^++H^+	$[H^+] = c_{HCl}$	4.3
22.00	110.0			2.32
40.00	200.0			1.48

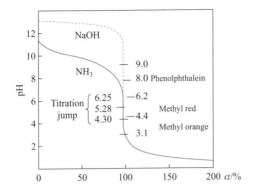

Fig.3-14 Titration curves of 20.00mL 0.1000mol·L^{-1} NH$_3$ titrated by 0.1000mol·L^{-1} HCl

图 3-14 用 0.1000 mol·L^{-1} HCl 溶液滴定 20.00mL 0.1000 mol·L^{-1} NH$_3$ 溶液的滴定曲线

When weak base is titrated by strong acid, and the equivalence point and the titration jump both fall within the weak acid ranges, Methyl red, bromocresol green and methyl orange can be selected as indicators. But when the titrant concentration reaches 0.1 mol·L^{-1}, methyl orange cannot be used as an indicator, otherwise the end point error will increase. When a strong acid titrates a weak base, and when the concentration of the base is constant, the larger the K_b is, the stronger the basicity becomes, and the larger the titration jump range has on the titration curve; Otherwise, the smaller the jump range has. The situation is similar to that of weak acid by strong base. **Therefore, when a strong acid titrates a weak base, only when $cK_b \geq 10^{-8}$, the weak base can be directly titrated with a standard acid solution.**

强酸滴定弱碱的化学计量点及滴定突跃都在弱酸性范围内，可选用甲基红、溴甲酚绿甲基橙为指示剂。在滴定剂浓度为 0.1mol·L^{-1} 情况下不能采用甲基橙为指示剂，否则终点误差将增大。强酸滴定弱碱时，当碱的浓度一定时，K_b 越大即碱性越强，滴定曲线上滴定突跃范围也越大；反之，突跃范围越小。与强碱滴定弱酸的情况相似。因此，强酸滴定弱碱时，只有当 $cK_b \geq 10^{-8}$，此弱碱才能用标准酸溶液直接滴定。

Section 7 Preparation and Calibration of Acid-base Standard Solution

第 7 节 酸碱标准液的配制和标定

1. The Primary Standard Substance

1. 基准物质

In titration analysis, hydrochloric acid and sulfuric acid solutions are commonly used as titrants (standard solutions), especially hydrochloric acid solution, because of its low price and easy availability, is a very common one. Diluted hydrochloric acid solution has strong

在滴定分析法中常用盐酸、硫酸溶液作为滴定剂（标准溶液），尤其是盐酸溶液，因其价格低廉，易于得到。稀盐酸溶液无氧化还原性质，酸性强且稳定，因此用得较多，但市售盐酸中

acidity and stability, and is free from redox, so they are used more. However, the HCl in hydrochloric acid sold at market is unstable and often contains impurities. HCl standard solution is usually prepared by an indirect method, which is first prepared into a solution of approximate concentration, and then calibrated with a primary standard substance. Calibration primary standard substances are commonly used with anhydrous Na_2CO_3 or borax ($Na_2B_4O_7 \cdot 10H_2O$) and other materials.

(1) Anhydrous Na_2CO_3 This substance is easy to absorb moisture in the air and it should be dried at 180-200℃ for 2-3 hours before use. $NaHCO_3$ also can be used. Firstly, it should be dried at 270-300℃ for one hour, then after drying, it will be decomposed and converted into Na_2CO_3 and then be stored in a dryer.

HCl含量不稳定，且常含有杂质，应采用间接法配制，再用基准物质标定，确定其准确浓度，常用无水Na_2CO_3或硼砂（$Na_2B_4O_7 \cdot 10H_2O$）等基准物质进行标定。

（1）无水Na_2CO_3　无水Na_2CO_3易吸收空气中的水分，故使用前应在180～200℃下干燥2～3h。也可用$NaHCO_3$在270～300℃下干燥1h，经烘干发生分解，转化为Na_2CO_3，然后放在干燥容器中保存。

Preparation of HCl solution
盐酸溶液的配制

$$2NaHCO_3 \rightleftharpoons Na_2CO_3 + CO_2\uparrow + H_2O$$

Suppose the concentration of hydrochloric acid to be calibrated is about 0.1 mol·L^{-1}, and the volume of hydrochloric acid to be consumed is 20-30 mL. According to the titration reaction, it can be calculated that the mass of Na_2CO_3 should be 0.11-0.16 g, and methyl orange can be used for titration as the indicator. When the solution turns from yellow to orange, it marks as the end point.

(2) Borax Borax is not easy to absorb water but easy to lose water, so it is required to be kept in a relative humidity of 40%-60% to ensure that the amount of crystalline water contained in it matches the chemical formula used in the calculation. At laboratories, people often use the method of filling a saturated aqueous solution of salt and sucrose at the bottom of the dryer to maintain the relative humidity at 60%.

The reaction of HCl calibrated by borax:

例如，欲标定的盐酸浓度约为0.1mol·L^{-1}，欲使消耗盐酸体积20～30mL，根据滴定反应可算出称取Na_2CO_3的质量应为0.11～0.16g，滴定时可采用甲基橙为指示剂，溶液由黄色变为橙色即为终点。

（2）硼砂　硼砂不易吸水，但易失水，因而要求保存在相对湿度为40%～60%的环境中，以确保其所含的结晶水数量与计算时的化学式相符。实验室常采用在干燥器底部装入食盐和蔗糖的饱和水溶液的方法，使相对湿度维持在60%。

硼砂标定盐酸的反应：

$$B_4O_7^{2-} + 5H_2O \longrightarrow 2H_3BO_3 + 2H_2BO_3^-$$

$$2H_2BO_3^- + 2HCl \longrightarrow 2H_3BO_3 + 2Cl^-$$

Overall reaction

总反应式为：

$$B_4O_7^{2-} + 5H_2O + 2HCl \longrightarrow 4H_3BO_3 + 2Cl^-$$

One molecular $B_4O_7^{2-}$ interacts with water to produce two molecular $H_2BO_3^-$ and two molecular H_3BO_3, of which only two molecular $H_2BO_3^-$ can be affected by HCl; Thus,

$$1B_4O_7^{2-} \to 2H_2BO_3^- \to 2H^+$$

Since the reaction product is H_3BO_3, if at the equivalence point it is $[H_3BO_3]=5.0\times10^{-2}$ mol·L^{-1}, and the K_a of H_3BO_3 is known, that is, $K_a=5.7\times10^{-10}$ mol·L^{-1}, the calculation formula of $[H^+]$ at the equivalence point is:

$$[H^+] = \sqrt{cK_a} = \sqrt{5.0\times10^{-2}\times5.7\times10^{-10}} = 5.3\times10^{-6}(mol\cdot L^{-1})$$

pH=5.28

During the titration, methyl red can be selected as the indicator, and the end point reaches when the solution turns from yellow to red.

Assuming that the concentration of hydrochloric acid to be calibrated is about 0.1 mol·L^{-1}, and the volume of hydrochloric acid to be consumed is 20-30 mL, it can be calculated that 0.38-0.57g of borax should be weighed. Because the molar mass of borax (381.4 g·mol^{-1}) is larger than Na_2CO_3, the mass of borax required to calibrate the same concentration of hydrochloric acid is also more than that of Na_2CO_3, which means the relative error of weighing becomes small. In other words, borax is superior to Na_2CO_3 as a primary standard substances for hydrochloric acid calibration.

Besides the abovementioned these two primary standard substances, $KHCO_3$, potassium hydrogen tartrate and others are also used to calibrate the hydrochloric acid solution.

2. Preparation and calibration of standard base solution

Standard base solutions are usually prepared with NaOH. Solid NaOH is not only easy to absorb moisture, but easy to absorb CO_2 and water, which produces a small amount of Na_2CO_3. NaOH may also contain impurities of sulfate, silicate, chloride and so on, so the standard solution should be prepared using the indirect method,

1分子 $B_4O_7^{2-}$ 与水作用产生2分子 $H_2BO_3^-$ 和2分子 H_3BO_3，其中仅有2分子 $H_2BO_3^-$ 能被 HCl 作用。

由于反应产物是 H_3BO_3，若化学计量点时，$[H_3BO_3]=5.0\times10^{-2}$ mol·L^{-1}，已知 H_3BO_3 的 $K_a=5.7\times10^{-10}$，则化学计量点时 $[H^+]=5.3\times10^{-6}$ mol·L^{-1}。

滴定时可选择甲基红为指示剂，溶液由黄色变为红色即为终点。

例如，待标定的盐酸浓度约为 0.1mol·L^{-1}，欲使消耗的盐酸体积为 20～30mL，可算出应称取硼砂的质量为 0.38～0.57g，由于硼砂的摩尔质量（381.4g·mol^{-1}）比 Na_2CO_3 大，标定同样浓度的盐酸所需的硼砂质量也比 Na_2CO_3 多，因而称量的相对误差就小，所以硼砂作为标定盐酸的基准物质优于 Na_2CO_3。

除上述两种基准物质外，还有 $KHCO_3$、酒石酸氢钾等基准物质可用于标定盐酸溶液。

2. 碱标准溶液的配制和标定

氢氧化钠是最常用的碱溶液。固体氢氧化钠具有很强的吸湿性，易吸收 CO_2 和水分，生成少量 Na_2CO_3，且含少量的硅酸盐、硫酸盐和氯化物等，因而不能直接配制成标准溶液，只能用间接法配制，再以基准物质标定其浓度。常用邻苯二甲酸氢钾基准物质标定。

namely with base solution approximate concentrate, then perform the calibration. Commonly used Potassium iphthalate is commonly used for calibration as primary standard substance.

Calibration of NaOH solution
NaOH 溶液标定

The molecular formula of potassium hydrogen iphthalate is $C_8H_4O_4HK$, and its structural formula is:

邻苯二甲酸氢钾的分子式为 $C_8H_4O_4HK$，其结构式为：

The molar mass of $C_8H_4O_4HK$ is 204.2 g·mol^{-1}. It belongs to a weak organic acid salt and shows acidity in aqueous solution. Because of $cK_{a2} \geqslant 10^{-8}$, it can be titrated with NaOH solution. The product of titration is potassium sodium phthalate, which can accept protons in aqueous solution, showing the nature of base.

$C_8H_4O_4HK$ 的摩尔质量为 204.2g·mol^{-1}，属有机弱酸盐，在水溶液中呈酸性，因 $cK_{a2} \geqslant 10^{-8}$，故可用 NaOH 溶液滴定。滴定的产物是邻苯二甲酸钾钠，它在水溶液中能接受质子，显示碱的性质。

Assuming that the concentration of $C_8H_4O_4HK$ at the beginning is 0.10 mol·L^{-1}. When the equivalence point is got at, the volume doubles and the concentration of potassium sodium phthalate reaches c=0.050 mol·L^{-1}. The pH at the equivalence point should be calculated as follows:

设邻苯二甲酸氢钾溶液开始时浓度为 0.10mol·L^{-1}，到达化学计量点时，体积增加一倍，邻苯二甲酸钾钠的浓度为 0.050mol·L^{-1}。化学计量点时 pH=9.11。

$$[OH^-] = \sqrt{cK_{b1}} = \sqrt{\frac{cK_w}{K_{a2}}} = \sqrt{\frac{0.05 \times 1.0 \times 10^{-14}}{2.9 \times 10^{-6}}} = 1.3 \times 10^{-5} \text{ mol·L}^{-1}$$

pOH=4.89，pH=9.11

At this time, the solution shows basicity, and phenolphthalein or thymol blue can be used as indicators. Besides $C_8H_4O_4HK$, primary standard substances such as oxalic acid, benzoic acid, and hydrazine sulfate ($N_2H_4·H_2SO_4$) all can be used to calibrate NaOH solution.

此时溶液呈碱性，可选用酚酞或百里酚蓝为指示剂。除邻苯二甲酸氢钾外，还有草酸、苯甲酸、硫酸肼（$N_2H_4·H_2SO_4$）等基准物质可用于标定 NaOH 溶液。

Experiment 2 Preparation and Standardization of Standard 0.1 mol·L^{-1} Hydrochloric acid Solution

1. Objective

(1) Master the principle and method of using anhydrous sodium carbonate as primary standard substance to standardize hydrochloric acid solution.
(2) Judge correctly end point of the mixed indicator of methyl red-bromocresol green solution.
(3) Master the operation of the acid burette.

2. Principle

The commercial hydrochloric acid is a colorless and transparent aqueous solution of hydrogen chloride. It is easy to volatize and contains about 36% to 38% (w/w) of HCl. So the standard solution can not be prepared directly, so indirect method is adopted for the preparation of 0.1 mol·L^{-1} hydrochloric acid. That is, an solution of the approximate concentration is firstly prepared, and then standardized with primary standard substance.

There are a number of primary standards used as standardization for hydrochloric acid, such as anhydrous sodium carbonate, borax, and so on. We use anhydrous sodium carbonate as primary standard and methyl red-bromocresol green mixed solution as indicator. The color changes from green to dark purple at the end point.

The reaction of titration may be represented in the following equation:

$$2HCl + Na_2CO_3 \longrightarrow 2NaCl + H_2CO_3 \qquad H_2CO_3 \longrightarrow H_2O + CO_2 \uparrow$$

3. Apparatus and Materials

(1) Apparatus cylinder (10 mL), glass reagent bottle (1000 mL), acid burette (25 mL), conical flask (250 mL), electronic analytical balance.
(2) Materials primary standard (anhydrous Na_2CO_3), concentrated hydrochloric acid (36%-38%), methyl red-bromocresol green indicator.

4. Procedures

(1) Preparation of 0.1 mol·L^{-1} hydrochloric acid solution Measure out 9 mL of concentrated hydrochloric acid by a small cylinder, and transfer to a clean glass reagent bottle. Dilute it to 1000 mL with distilled water and thoroughly mix by shaking.
(2) Standardization of 0.1 mol·L^{-1} hydrochloric acid with anhydrous sodium carbonate Weigh out accurately about 0.12 g of primary standard anhydrous sodium carbonate (which has been previously dried at a temperature of about 270 ℃ to 300 ℃ until the weight is constant). Dissolve it with 25 mL of distilled water, and add 4-5 drops of methyl red-bromocresol green solution. Titrate it with 0.1 mol·L^{-1} hydrochloric acid until the color changes from green to purple-red. Boil the solution for about two minutes, then cool it to room temperature and continue the titration until the color of solution changes from green to dark purple. Set down the reading. Repeat the

standardization for another 2 times.

The concentration of hydrochloric acid can be calculated from the formula:

$$c_{HCl} = \frac{w_{Na_2CO_3}}{V_{HCl} \times \dfrac{M_{Na_2CO_3}}{2000}} \qquad M_{Na_2CO_3} = 106.0 \text{ g} \cdot \text{mol}^{-1}$$

5. Notes

(1) Na_2CO_3 is easy to absorb water and the weighing process is required rapid.

(2) pH of solution does not change greatly when the titration is close to the end point due to the forming of H_2CO_3-$NaHCO_3$ buffer solution. So we have to boil the solution for 2 minutes to break the buffer solution, and then cool it to room temperature and continue the titration.

6. Questions

(1) How to prepare 1000 mL 0.1 mol·L^{-1} hydrochloric acid solution.

(2) Whether the conical flask must be dried and the quantity of distilled water must be accurate during the experiment.

(3) Why is the solution boiled when titration is nearing completion? Why is the the solution cooled to room temperature?

(4) If 23 mL 0.1 mol·L^{-1} hydrochloric acid is required for standardization of hydrochloric acid in this experiment, how many grams of anhydrous sodium carbonate must be weighed out?

实验2　0.1 mol·L^{-1} HCl 标准溶液的配制和标定

1. 实验目的

（1）掌握用无水碳酸钠作为基准物质标定盐酸溶液的原理和方法。

（2）正确判断混合指示剂甲基红-溴甲酚绿的滴定终点。

（3）掌握酸式滴定管的操作。

2. 实验原理

市售盐酸是一种无色透明的氯化氢水溶液，容易挥发，大约含36%～38%（质量分数）的 HCl，因此不能直接配制。可采用间接法配制 0.1 mol·L^{-1} 盐酸溶液，首先配制一个大约浓度的溶液，然后用基准物质标定。

许多基准物质可以用于 HCl 溶液的标定，如无水碳酸钠、硼砂等。我们用无水碳酸钠作为基准物质，用甲基红-溴甲酚绿混合溶液作为指示剂，滴定终点颜色从绿色变为深紫色。

滴定反应可用下式表示：

$$2HCl + Na_2CO_3 \longrightarrow 2NaCl + H_2CO_3 \qquad H_2CO_3 \longrightarrow H_2O + CO_2 \uparrow$$

3. 仪器和材料

（1）仪器　量筒（10mL），试剂瓶（1000mL），酸式滴定管（25mL），锥形瓶（250mL），电子分析天平。

（2）材料　无水 Na_2CO_3 基准物质，浓 HCl（36%～38%），甲基红-溴甲酚绿试剂。

4. 实验步骤

（1）$0.1mol \cdot L^{-1}$ HCl 溶液的配制　用小量筒取 9mL 浓 HCl 转移至洁净带塞玻璃瓶中，用蒸馏水稀释至 1000mL 并振摇使之混匀。

（2）用无水 Na_2CO_3 标定 $0.1mol \cdot L^{-1}$ HCl 溶液　精确称取在 270～300℃下干燥至恒重的无水 Na_2CO_3 基准物质 0.12g，用 25mL 蒸馏水溶解，加入 4～5 滴甲基红-溴甲酚绿溶液，用 $0.1mol \cdot L^{-1}$ HCl 滴定至颜色从绿色变为紫红色。煮沸溶液两分钟后冷却至室温，并继续滴定直至溶液由绿色变为深紫色。记下读数。重复滴定两次。

HCl 溶液浓度可由以下公式计算：$c_{HCl} = \dfrac{w_{Na_2CO_3}}{V_{HCl} \times \dfrac{M_{Na_2CO_3}}{2000}}$

5. 注意事项

（1）Na_2CO_3 易吸水，称量速度要快。

（2）pH 在终点附近变化不显著，终点变化不敏锐，因为形成了 Na_2CO_3-$NaHCO_3$ 缓冲溶液。所以必须煮沸 2min 来破坏缓冲溶液，然后冷却至室温，再继续滴定。

6. 思考

（1）如何配制 1000mL $0.1mol \cdot L^{-1}$ 盐酸溶液？

（2）锥形瓶是否需要干燥？实验中蒸馏水的用量是否需要精确？

（3）当滴定接近完成时，为什么溶液要煮沸？为什么溶液要冷却到室温？

（4）在这个实验中如果需要消耗 23mL $0.1mol \cdot L^{-1}$ 盐酸标准溶液，要称量多少无水 Na_2CO_3？

Experiment 3　Preparation and Standardization of Standard Solution of 0.1 mol·L^{-1} Sodium Hydroxide Solution

1. Objective

(1) Master how to prepare and standardize sodium hydroxide standard solution with potassium acid phthalate primary standard substances.

(2) Master the operation of the basic burette.

(3) Learn to judge the end point of phenolphthalein correctly.

2. Principle

Sodium hydroxide can rapidly absorb water and carbon dioxide from the air. A certain amount of sodium carbonate is always present.

$$2NaOH + CO_2 \longrightarrow Na_2CO_3 + H_2O$$

So indirect method is adopted for the preparation of 0.1 mol/L sodium hydroxide solution. It is customary to prepare sodium hydroxide solution of approximate concentration and then standardize it against a primary standard to obtain the exact concentration.

There are several methods can be used to prepare carbonate-free standard sodium hydroxide solution, and the saturated sodium hydroxide solution method is thought to the most common method. Sodium carbonate is insoluble in the saturated sodium hydroxide solution, and can be precipitated to the bottom of the bottle. Siphon off the clear supernatant liquid and dilute it with carbon dioxide-free water to the approximate concentration, then standardize the standard solution against a primary standard.

There are several primary standard acids to standardize the basic solution, for example, oxalic acid dihydrate($H_2C_2O_4 \cdot 2H_2O$), benzoic acid (C_6H_5COOH), sulfamic acid (NH_2SO_3H) and potassium acid phthalate ($KHC_8H_4O_4$) etc. At present the last one is preferable, and the reaction is as follows:

$$\text{C}_6\text{H}_4(\text{COOH})(\text{COOK}) + NaOH \longrightarrow \text{C}_6\text{H}_4(\text{COONa})(\text{COOK}) + H_2O$$

Phenolphthalein solution is used as indicator, the solution at the equivalence point would be slightly alkaline, due to the hydrolysis of the weak acid salt.

3. Apparatus and Materials

(1) Apparatus   basic burette (25 mL), conical flask (250 mL), cylinder (100 mL), beaker (500 mL), plastic reagent bottle (500 mL), electronic analytical balance

(2) Materials   NaOH (A. R.), phenolphthalein indicator.

4. Procedures

(1) Preparation of saturated NaOH solution   Dissolve 120g of NaOH in 100mL of distilled water, shake well, cool the solution and transfer to a plastic reagent bottle. Allow the solution to stand for several days so that any Na_2CO_3 (which is insoluble in concentrated NaOH) will settle to the bottom.

(2) Preparation of 0.1 mol $\cdot$ L^{-1} NaOH solution   Siphon off 2.8 mL of the saturated NaOH solution to a plastic reagent bottle. Add distilled water which is freshly boiled and cooled to make 500mL, and shake well.

(3) Standardization of 0.1 mol $\cdot$ L^{-1} NaOH solution   Weigh out accurately about 0.45 g primary standard potassium acid phthalate (dried at 105-110℃ until the weight is constant) to a conical flask, adding 30 mL of distilled water which is freshly boiled and cooled to the flask and shake gently until the sample is dissolved Add 2 drops of phenolphthalein indicator solution, titrate the solution with 0.1 mol $\cdot$ L^{-1} NaOH solution to the first permanent pink color, which should persist not less than thirty seconds. Record the final reading. Repeat the standardization twice.

Calculate the concentration of the NaOH standard solution based on the weight of the salt and the final reading of the NaOH.

$$c_{NaOH} = \frac{w_{KHC_8H_4O_4} \times 1000}{V_{NaOH} \times M_{KHC_8H_4O_4}} \qquad M_{KHC_8H_4O_4} = 204.2 \text{g·mol}^{-1}$$

5. Notes

(1) We must weigh sodium hydroxide in a beaker instead of on a piece of paper.
(2) Adjust liquid level of the burette to zero point before each titration.

6. Questions

(1) How many grams of potassium acid phthalate is needed if 23 mL of 0.1 mol·L^{-1} NaOH is required to neutralize it.
(2) If the primary standard potassium acid phthalate has not been previously dried at a temperature of about 105℃ to 110℃, the concentration of the NaOH standard solution will be higher or lower?

实验 3　0.1mol·L^{-1} 氢氧化钠标准溶液的配制与标定

1. 实验目的

（1）掌握氢氧化钠溶液配制方法及用基准物质邻苯二甲酸氢钾标定标准溶液的方法。
（2）掌握碱式滴定管的操作方法。
（3）学会正确判断酚酞的滴定终点。

2. 实验原理

NaOH 会迅速从空气中吸 H$_2$O 和 CO$_2$，一定量的 Na$_2$CO$_3$ 总是存在：

$$2NaOH + CO_2 \longrightarrow Na_2CO_3 + H_2O$$

所以配制 0.1mol·L^{-1} NaOH 溶液需采用间接配制法。通常先配制近似浓度的 NaOH 溶液，然后用基准物质标定得其准确浓度。

配制无碳酸盐的标准 NaOH 溶液方法很多，最常用的是用饱和 NaOH 溶液来配制。碳酸钠在饱和 NaOH 溶液中不溶解而沉淀下来，吸取上清液并用无 CO$_2$ 的水稀释成近似所需浓度的溶液，再用基准物质标定。有许多基准酸可用来标定碱性溶液，如含 2 个结晶水的草酸（H$_2$C$_2$O$_4$·2H$_2$O）、苯甲酸（C$_6$H$_5$COOH）、氨基磺酸（NH$_2$SO$_3$H）和邻苯二甲酸氢钾（KHC$_8$H$_4$O$_4$）等。目前，邻苯二甲酸氢钾最常用，其反应如下：

C$_6$H$_4$(COOH)(COOK) + NaOH ⟶ C$_6$H$_4$(COONa)(COOK) + H$_2$O

用酚酞溶液作为指示剂，在滴定终点由于弱酸盐的水解，溶液会略显碱性。

3. 仪器和材料

（1）仪器　碱式滴定管（25mL），锥形瓶（250mL），量筒（100mL），烧杯（500mL），塑料试剂瓶（500mL），电子分析天平。

（2）材料　氢氧化钠（分析纯），酚酞指示剂。

4. 实验步骤

（1）饱和氢氧化钠溶液的配制　在100mL蒸馏水中溶解120g NaOH，充分摇匀，冷却溶液并转移到塑料瓶内。放置一定时间，使所有不能溶解在浓NaOH溶液中的Na_2CO_3沉淀至底部。

（2）0.1mol·L^{-1} NaOH溶液的配制　吸取2.8mL的饱和NaOH溶液至塑料试剂瓶，加入新沸并放冷的蒸馏水至500mL，充分振摇。

（3）0.1mol·L^{-1} NaOH溶液的标定　称取约0.45g基准物质邻苯二甲酸氢钾（在105～110℃干燥至恒重），精确至±0.0001g，每个锥形瓶中加入30mL新沸过并放冷的蒸馏水，轻轻振摇至样品全部溶解，加入2滴酚酞指示剂，用0.1mol·L^{-1} NaOH溶液滴定至粉色，并保持30s不褪色。记录最终读数。重复2次。

根据称量的邻苯二甲酸氢钾的质量和消耗的NaOH量，计算NaOH标准溶液的浓度。

$$c_{NaOH} = \frac{w_{KHC_8H_4O_4} \times 1000}{V_{NaOH} \times M_{KHC_8H_4O_4}}$$

5. 注意事项

（1）氢氧化钠必须在烧杯中称量，不能在称量纸上称量。

（2）每次滴定前，滴定管的液面要调到零。

6. 思考

（1）如果滴定23mL 0.1mol·L^{-1} NaOH溶液，需要多少克邻苯二甲酸氢钾？

（2）如果基准物质邻苯二甲酸氢钾没有预先在105～110℃中干燥，则NaOH标准溶液的浓度会有何影响？

Exercises

3-1　The conjugate base of $H_2PO_4^-$ is (　).

　　A. HPO_4^{2-}　　　　B. PO_4^{3-}　　　　C. H_3PO_4　　　　D. OH^-

3-2　In the following acid-base pairs, (　) is the conjugate acid-base pair.

　　A. H_2CO_3-CO_3^{2-}　　B. H_3O^+-OH^-　　C. HPO_4^{2-}-PO_4^{3-}　　D. H^+-OH^-

3-3　If the concentration is the same, which of the following ions has the strongest alkalinity in aqueous solution?

　　A. CN^- ($K_{CN^-}=6.2\times10^{-10}$)　　　　B. S^{2-} (H_2S: $K_{a1}=1.3\times10^{-7}$, $K_{a2}=7.1\times10^{-15}$)

　　C. F^- ($K_{HF}=3.5\times10^{-4}$)　　　　　D. CH_3COO^- ($K_{HAc}=1.8\times10^{-5}$)

3-4 When the concentration of HAc is x mol·L^{-1} and pH = pK_a, $\delta_{HAc}+\delta_{Ac^-}=$ ( ).
   A. x　　　　　B. 0.5　　　　　C. 1　　　　　D. uncertain

3-5 In H$_2$A solution, when pK_{a1}>pH>pK_{a2}, the main type is ( ).
   A. H$_2$A　　　B. HA$^-$　　　C. A^{2-}　　　D. uncertain

3-6 When two hydrochloric acids with pH=5 and pH=3 are mixed in a volume ratio of 1+2, the pH of the mixed solution is ( ).
   A. 3.17　　　　B. 10.1　　　　C. 5.3　　　　D. 8.2

3-7 Add ( ) water to 1 mL of hydrochloric acid with pH= 1.8 to change the solution to pH=2.8.
   A. 9mL　　　　B. 10mL　　　　C. 8mL　　　　D. 12mL

3-8 It is known that K_{HAc}=1.75×10^{-5}, the pH of 0.20 mol·L^{-1} NaAc solution pH is ( ).
   A. 2.72　　　　B. 4.97　　　　C. 9.03　　　　D. 11.27

3-9 The pH of 0.20 mol·L^{-1} H$_3$PO$_4$ solution is ( ). (K_{a1}=7.6×10^{-3}, K_{a2}=6.3×10^{-8}, K_{a3}=4.4×10^{-13})
   A. 1.41　　　　B. 1.45　　　　C. 1.82　　　　D. 2.82

3-10 The minimum change in pH of the following solution diluted by 10 times is ( ).
   A. 1mol·L^{-1} HAc　　　　　　　B. 1mol·L^{-1} HAc 和 0.5mol·L^{-1} NaAc
   C. 1mol·L^{-1} NH$_3$　　　　　　D. 1mol·L^{-1} NH$_4$Cl

3-11 The pH of the buffer solution obtained by mixing 100 mL 0.30mol·L^{-1} NH$_3$·H$_2$O with 100mL 0.45mol·L^{-1} NH$_4$Cl is ( ). K_b of ammonia is 1.8×10^{-5} (assuming that the total volume after mixing is the sum of the volumes before mixing)
   A. 11.85　　　B. 6.78　　　　C. 9.08　　　　D. 13.74

3-12 The buffer range of the buffer solution is ( ).
   A. pH=pK_a+1　　B. pH=pK_a−1　　C. pH=pK_a±1　　D. pH=pK_a±2

3-13 The correct range of color changes for the following indicators is ( ).
   A. methyl red（3.8-4.4）　　　　B. phenolphthalein（8.0-9.8）
   C. thymolphthalein（8.0-10.0）　　D. methyl orange（3.1-4.4）

3-14 When the concentration is constant, titration curves are obtained by titrating different unary weak acids with strong base, and the titration jump changes with K_a value as follows: ( ).
   A. The larger the K_a is, the larger the jump range is
   B. The larger the K_a is, the smaller the jump range is
   C. The change of jump range is not affected by K_a
   D. Uncertain

3-15 Titrate 0.1 mol·L^{-1} a weak monoprotic base (pK_b=5.0) with 0.10 mol·L^{-1} hydrochloric acid solution, the pH jump range is ( ).
   A. 6.7～5.3　　B. 6.7～4.3　　C. 7.7～5.3　　D. 5.7～4.3

3-16 When anhydrous Na$_2$CO$_3$ is used to calibrate HCl solution, if bromocresol green-methyl red is used as an indicator, the end point color change is that ( ).
   A. green turns dark red　　　　B. blue turns yellow
   C. yellow turns red　　　　　　D. yellow turns green

3-17　The standard substance for calibrating NaOH standard solution is ( ).
　　　A. anhydrous sodium carbonate　　　B. $Na_2B_4O_7 \cdot 10H_2O$
　　　C. potassium acid phthalate　　　　D. potassium acid tartrate

3-18　Pick out which of the following substances are bases, acids or amphiprotic.
　　　H_2SO_3; $KHC_8H_4O_4$; H_2O; NH_4^+; $Na_2B_4O_7$; CH_3COOH; Na_2CO_3; $NaHCO_3$; CH_3COO^-; NaOH.

3-19　The dissociation of polyacid in aqueous solution is graded and there are multiple conjugate acid-base pairs, then what are the conjugate acid-base pairs in the dissociation process of H_3PO_4?

3-20　What kind of solution is a buffer?

3-21　What are the four international standard buffer solutions ?

3-22　What are the factors that affect the size of the titration jump when an acid is titrated by a strong base?

3-23　Calculate δ_{Ac^-} and δ_{HAc} in HAc solution at pH=5. ($K_a = 1.8 \times 10^{-5}$)
　　　　　　　　　　　　　　　　　　　　　　　　　　　　(δ_{Ac^-}=0.64 and δ_{HAc}=0.36)

3-24　When the same volume of hydrochloric acid solution of pH=5 is mixed with sodium hydroxide solution of pH=12, what is the pH of the solution?
　　　　　　　　　　　　　　　　　　　　　　　　　　　　(pH=11.7)

3-25　What is the pH of 0.20 mol·L^{-1} NaAc solution ? $K_a = 1.8 \times 10^{-5}$.
　　　　　　　　　　　　　　　　　　　　　　　　　　　　(pH=9.03)

3-26　How many mL of 0.500 mol·L^{-1} NH_4Cl and 0.500 mol·L^{-1} $NH_3 \cdot H_2O$ should be taken to prepare 200mL of buffer solution with pH=8.00? K_b of ammonia is 1.8×10^{-5}. (Assume that the total volume after mixing is the sum of the volumes before mixing.)
　　　　　　　　　　　　　　　　　　　　　　(V_{NH_4Cl}=189mL, $V_{NH_3 \cdot H_2O}$=11mL)

3-27　Titrate 20.00mL 0.1000 mol·L^{-1} $NH_3 \cdot H_2O$ with 0.1000 mol·L^{-1} HCl, calculate the pH of the solution at stoichiometric point.
　　　　　　　　　　　　　　　　　　　　　　　　　　　　(pH=5.28)

Chapter 4 Coordination Titration
第 4 章 配位滴定法

 Study Guide 学习指南

The knowledge objectives to be mastered in this chapter are as follows: To comprehend the properties of ethylene diamine tetra acetic acid (EDTA), and the characteristics of its forming complexes with metal ions. To master the main reaction between EDTA and metal ions, the main side reaction and how to calculate the side reaction coefficient and the conditional formation constant. To comprehend the basic principles of complexometric titrations. To master the method of making acidic effective curve. To understand the working principles of metal indicators. To master the principles of improving the selectivity of complexometric titration. The competence objectives to be mastered in this chapter are as follows: To calculate the minimum pH and suitable pH ranges for titrating different metal ions accurately. To use metal indicator correctly. To explain and apply methods to improve the selectivity of complexometric titration by combining with analysis and practice. To choose different complexometric titration methods to determine different metal ions.

通过本章的学习，掌握乙二胺四乙酸（简称 EDTA）的性质、它与金属离子形成配合物的特点；掌握 EDTA 与金属离子的主反应、主要的副反应和副反应系数的计算方法、条件稳定常数的计算方法；掌握配位滴定法的基本原理，掌握酸效应曲线的制作方法，了解金属指示剂的作用原理及掌握提高配位滴定选择性的原理；能正确计算滴定不同金属离子的最小 pH、适宜的 pH 范围，能正确使用金属指示剂，能结合分析与实践来解释、应用提高配位滴定选择性的方法，能合理选择不同的配位滴定方式来测定不同的金属离子。

Section 1 EDTA and Its Complexes
第 1 节 EDTA 的性质及其配合物

Coordination titration is a titration analysis method based on complexing reaction, also known as complexometric titration. Complexing reactions are very

配位滴定法是以配位反应为基础的滴定分析方法，亦称络合滴定法。在化学反应中，配位反应是非常普遍的。但

common in chemical reactions. However, before the aminocarboxyl ligand was used in analytical chemistry in 1945, the application of the complexometric titration was very limited, which attributed to the fact that many electrodeless complexes are not stable enough to meet the requirements of the titration reaction; in other words, because stepwise coordination comes out during the complexometric process and the stability constants at various levels are not much different, the end point of the titration is not obvious. Since aminocarboxyl ligand was introduced into titration analysis, the complexometric titration has been developed rapidly.

As a kind of organic ligand containing aminodiacetic acid group [—N(CH$_2$COOH)$_2$] as the host, aminocarboxyl ligand contains two kinds of ligating atoms of ammonia nitrogen with strong complexometric ability, which can interact with most metal ions to form stable certain complexes. There are many types of aminocarboxyl ligands, the followings are some important ones:

在1945年将氨羧配体用于分析化学以前，配位滴定法的应用非常有限，这是由于：许多无机配合物不够稳定，不符合滴定反应的要求；在配位过程中有逐级配位产生，各级稳定常数相差又不大，以至滴定终点不明显。自从滴定分析中引入了氨羧配体之后，配位滴定法才得到了迅速的发展。

氨羧配体是一类含有以氨基二乙酸基团［—N(CH$_2$COOH)$_2$］为基体的有机配位体，它含有配位能力很强的氨氮两种配位原子，能与多数金属离子形成稳定的而且组成一定的配合物。氨羧配体的种类有很多，比较重要的有

Ethylene diamine tetra acetic acid (乙二胺四乙酸，EDTA)

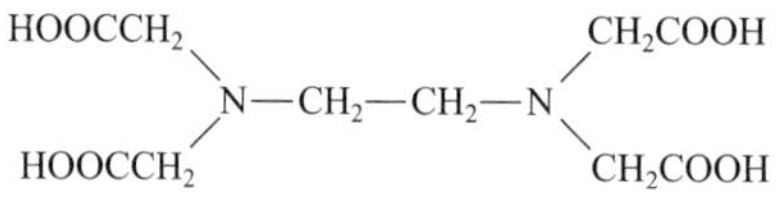

$$\text{HOOCCH}_2\diagdown_{\text{HOOCCH}_2}\!\!\text{N—CH}_2\text{—CH}_2\text{—N}\diagup^{\text{CH}_2\text{COOH}}_{\text{CH}_2\text{COOH}}$$

Cycloethane diamine tetra acetic acid (环乙烷二胺四乙酸，CDTA or DCTA)

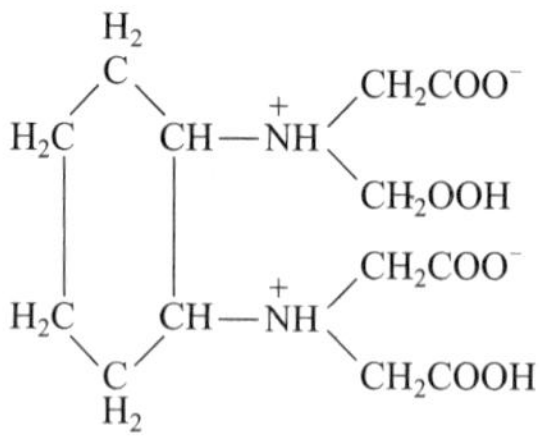

Egtazic acid, Glycol ether diamine tetra acetic acid (乙二醇二乙醚二胺四乙酸，EGTA)

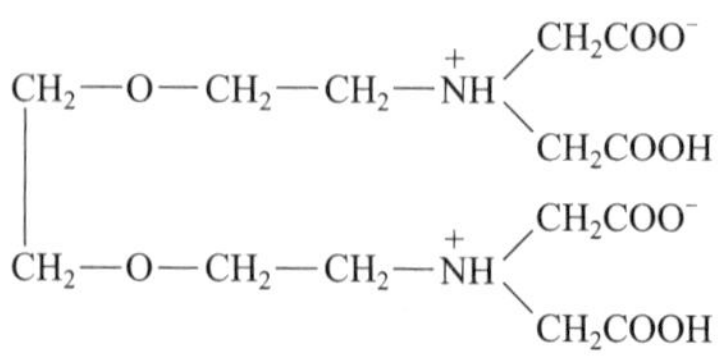

Ethylenediamine tetra propionic acid (乙二胺四丙酸，EDTP)

$$\begin{array}{c} CH_2-\overset{+}{N}H \begin{array}{l} CH_2COO^- \\ CH_2COOH \end{array} \\ | \\ CH_2-\overset{+}{N}H \begin{array}{l} CH_2COO^- \\ CH_2COOH \end{array} \end{array}$$

In the complexometric titration, EDTA is the most important one.

在配位滴定中，EDTA 是最重要的一种。

1. EDTA and Its Disodium Salt

1. 乙二胺四乙酸及其二钠盐

Ethylene diamine tetra acetic acid (EDTA) is a tetraprotic acid and customarily signified by H_4Y. Because of its low solubility in water (at 22℃, it can only dissolve 0.02 g per 100 mL of water), its disodium salt $Na_2H_2Y \cdot 2H_2O$ is commonly used, also referred to as EDTA. The latter has high solubility (at 22℃, it can dissolve 11.1 g per 100 mL of water) and the concentration of its saturated aqueous solution is about 0.3 mol·L^{-1}. In aqueous solution, ethylene diamine tetra acetic acid has a double dipole ion structure:

乙二胺四乙酸（简称 EDTA）是一种四元酸。习惯上用 H_4Y 表示。由于它在水中的溶解度很小（在 22℃时，每 100 毫升水中仅能溶解 0.02g），故常用它的二钠盐 $Na_2H_2Y \cdot 2H_2O$，一般也简称 EDTA。后者的溶解度大（在 22℃时，每 100mL 水中能溶解 11.1g），其饱和水溶液的浓度约为 0.3mol·L^{-1}。在水溶液中，乙二胺四乙酸具有双偶极离子结构。

$$\begin{array}{c} HOOCCH_2 \\ ^-OOCCH_2 \end{array} \overset{+}{\underset{H}{N}}-CH_2-CH_2-\overset{+}{\underset{H}{N}} \begin{array}{c} CH_2COO^- \\ CH_2COOH \end{array}$$

In addition, two carboxylates can also accept protons. When the acidity is high, EDTA will be transformed into hexatomic acid H_6Y^{2+}. There is a series of dissociation equilibriums in aqueous solution as follows:

此外，两个羧酸根还可以接受质子，当酸度很高时，EDTA 便转变成为六元酸 H_6Y^{2+}，在水溶液中存在着以下一系列的解离平衡。

$H_6Y^{2+} \rightleftharpoons H^+ + H_5Y^+$ $K_{a1} = \dfrac{[H^+][H_5Y^+]}{[H_6Y^{2+}]} = 10^{-0.9}$

$H_5Y^+ \rightleftharpoons H^+ + H_4Y$ $K_{a2} = \dfrac{[H^+][H_4Y]}{[H_5Y^+]} = 10^{-1.6}$

$H_4Y \rightleftharpoons H^+ + H_3Y^-$ $K_{a3} = \dfrac{[H^+][H_3Y^-]}{[H_4Y]} = 10^{-2.0}$

$H_3Y^- \rightleftharpoons H^+ + H_2Y^{2-}$ $K_{a4} = \dfrac{[H^+][H_2Y^{2-}]}{[H_3Y^-]} = 10^{-2.67}$

$H_2Y^{2-} \rightleftharpoons H^+ + HY^{3-}$ $\qquad K_{a5} = \dfrac{[H^+][HY^{3-}]}{[H_2Y^{2-}]} = 10^{-6.16}$

$HY^{3-} \rightleftharpoons H^+ + Y^{4-}$ $\qquad K_{a6} = \dfrac{[H^+][Y^{4-}]}{[HY^{3-}]} = 10^{-10.26}$

It can be seen that EDTA exists in aqueous solution with seven species H_6Y^{2+}, H_5Y^+, H_4Y, H_3Y^-, H_2Y^{2-}, HY^{3-}, Y^{4-}. When the pH changes, the distribution fractions δ occupied by the various species are different. Based on the calculation, the distribution curves of various species of EDTA solution at different pH can be drawn, as shown in figure 4-1.

At different pH values, the main species of EDTA are listed in table 4-1. Among its species, only Y^{4-} can ligate directly with metal ions. Therefore, the lower the acidity of the solution is, the higher the distribution fraction of Y^{4-} is, implying the stronger the coordination ability of EDTA has.

可见，EDTA 在水溶液中以 H_6Y^{2+}、H_5Y^+、H_4Y、H_3Y^-、H_2Y^{2-}、HY^{3-}、Y^{4-} 七种形体存在，当 pH 不同时，各种存在型体所占的分布分数 δ 是不同的。根据计算，可以绘制不同 pH 时 EDTA 溶液中各种型体的分布曲线，如图 4-1 所示。

在不同 pH 时，EDTA 的主要存在形态列于表 4-1 中，在其中形态中，只有 Y^{4-} 能与金属离子直接配位。所以溶液的酸度越低，Y^{4-} 的分布数越大，EDTA 的配位能力越强。

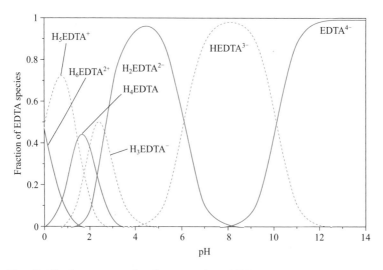

Fig.4-1 The distribution curves of various species of EDTA solution at different pH values
图 4-1 不同 pH 值下 EDTA 的不同形态的分布曲线

Table 4-1 The main species of EDTA at different pH values
表 4-1 不同 pH 值下 EDTA 的主要形态

pH	<1	1～1.6	1.6～2	2～2.7	2.7～6.2	6.2～10.3	>10.3
Main species	H_6Y^{2+}	H_5Y^{2+}	H_4Y	H_3Y^-	H_2Y^{2-}	HY^{3-}	Y^{4-}

2. EDTA and Its Metal Ion Complexes

EDTA molecule has two ammonia nitrogen atoms and four carboxyl oxygen atoms, all having a lone pair of electrons, namely, 6 ligating atoms in total. Therefore, most metal ions can form multiple five-membered rings with EDTA. For example, the structure of the complex between EDTA and Ca^{2+}, Fe^{3+} is shown in figure 4-2.

2. EDTA 与金属离子的配合物

EDTA 分子具有两个氨氮原子和四个羧氧原子，都有孤对电子，即有6个配位原子。因此，绝大多数的金属离子均能与 EDTA 形成多个五元环，例如 EDTA 与 Ca^{2+}、Fe^{3+} 的配合物的结构如图 4-2 所示。

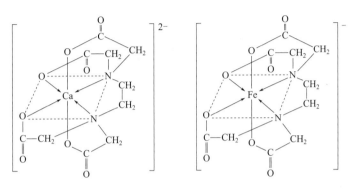

Fig.4-2 The structure of the complex between EDTA and Ca^{2+}, Fe^{3+}

图 4-2 EDTA 与 Ca^{2+}、Fe^{3+} 配合物的结构

Fig.4-2 shows that EDTA and metal ions form five five-membered rings, four ⌈—M—⌉ N—C—C—N five-membered rings and one ⌈—M—⌉ N—C—C—N five-membered ring. It is said that chelate with such kind of ring structure is very stable. Since the coordination number of most metal ions does not exceed 6, EDTA can form 1∶1 complexes with most metal ions, except for very few metal ions, such as zirconium (IV) and molybdenum (VI). When the colorless metal ions are coordinated with EDTA, a colorless integrated substance is formed. However, when the colored metal ions are coordinated with EDTA, a darker chelate is generally produced.

可以看出，EDTA 与金属离子形成五个五元环：四个 O-C-C-N 五元环及一个 N-C-C-N 五元环，具有这类环状结构的螯合物是很稳定的。由于多数金属离子的配位数不超过6，所以 EDTA 与大多数金属离子可形成 1∶1 型的配合物，只有极少数金属离子，如锆（IV）和钼（VI）等例外。无色的金属离子与 EDTA 配位时，则形成无色的螯合物，有色的金属离子与 EDTA 配位时，一般则形成颜色更深的螯合物。

NiY^{2-}	CuY^{2-}	CoY^{2-}	MnY^{2-}	CrY^-	FeY^-
Blue	Deep blue	Mauve	Mauve	Dark violet	Yellow

From the above, the chelates formed by EDTA and most metal ions have the following characteristics: ① The stoichiometric relation is simple, and there generally

综上所述，EDTA 与绝大多数金属离子形成的螯合物具有下列特点：①计量关系简单，一般不存在逐级配位现象。

exists no stepwise coordination. ② The complex is very stable and has excellent water solubility, so that the complexometric titration can be carried out in aqueous solution. These characteristics make EDTA as a titrant fully meet the requirements of analysis and determination, and that's why it is widely used.

②配合物十分稳定，且水溶性极好，使配位滴定可以在水溶液中进行。这些特点使EDTA滴定剂完全符合分析测定的要求，因而被广泛使用。

Section 2　Coordination Dissociation Equilibrium and Influencing Factors
第2节　配位解离平衡及影响因素

1. The Main Reaction Between EDTA and Metal Ions and the Stability Constants of the Complexes

1. EDTA与金属离子的主反应及配合物的稳定常数

EDTA can form 1∶1 complexes with most metal ions. The general reaction formula is as follows:

EDTA与金属离子大多形成1∶1型的配合物，下式为配位滴定的主反应。

$$M^{n+} + Y^{4-} \rightleftharpoons MY^{n-4}$$

The charge number of the ion can be omitted when writing, then it is abbreviated as:

书写时，可省略离子电荷数，简写为：

$$M + Y \rightleftharpoons MY \tag{4.1}$$

This reaction is the main reaction in complexometric titration. The stability constant of the complex at equilibrium is:

该反应是配位滴定的主反应。配合物的平衡稳定常数是：

$$K_{MY} = \frac{[MY]}{[M][Y]} \tag{4.2}$$

The stability constants of the complexes formed by common metal ions and EDTA are listed in table 4-2. Table 4-2 shows that when the stability of the metal ion and EDTA complexes varies greatly with the different kinds of the metal ions. Complexes of the alkali metal ion are the most unstable ones, $\lg K_{MY}$ remains at 2～3; For alkaline earth metal, $\lg K_{MY}$ is in 8～11; For divalent and transition metal, rare earth elements and Al^{3+}, $\lg K_{MY}$ is in 15～19; For trivalent and tetravalent metal and Hg^{2+}, $\lg K_{MY}>20$. The differences in stability of these complexes are mainly dependent on the ion charge

常见金属离子与EDTA所形成的配合物的稳定常数列于表4-2中。

从表4-2可以看出，金属离子与EDTA配合物的稳定性随金属离子的不同而差别较大。碱金属离子的配合物最不稳定，$\lg K_{MY}$在2～3；碱土金属离子的配合物，$\lg K_{MY}$在8～11；二价及过渡金属离子稀土元素及Al^{3+}的配合物，$\lg K_{MY}$在15～19；三价、四价金属离子和Hg^{2+}的配合物，$\lg K_{MY}>20$。这些配合物的稳定性的差别，主要取决于

Table 4-2 Stability constants of complexes by EDTA with some common metal ions
(Ion strength:I≈0.1, at 20℃)

表 4-2　EDTA 与几种常见金属离子配合物的稳定常数（20℃时，离子强度 ≈ 0.1）

Positive ion	$\lg K_{MY}$	Positive ion	$\lg K_{MY}$	Positive ion	$\lg K_{MY}$
Na^{2+}	1.66	Ce^{3+}	15.98	Cu^{2+}	18.80
Li^+	2.79	Al^{3+}	16.3	Hg^{2+}	21.8
Ba^{2+}	7.86	Co^{2+}	16.31	Th^{4+}	23.2
Sr^{2+}	8.73	Cd^{2+}	16.46	Cr^{3+}	23.4
Mg^{2+}	8.69	Zn^{2+}	16.50	Fe^{3+}	25.1
Ca^{2+}	10.69	Pb^{2+}	18.04	U^{4+}	25.80
Mn^{2+}	13.87	Y^{3+}	18.09	Bi^{3+}	27.94
Fe^{2+}	14.32	Ni^{2+}	18.62		

numbers, ion radius and electronic layer structure of the metal ion itself. In other words, the higher the ion charges are, the larger the ion radius is, and the more complex the structure of the electron shell is, the greater the stability constant of the complex remains. These are the essential factors that affect the stability of the complexes in terms of metal ions. In addition, changes in external conditions such as the acidity of the solution, temperature and the presence of other ligands also have an effect on the stability of the complexes.

金属离子本身的离子电荷数、离子半径和电子层结构。离子电荷数越高，离子半径越大，电子层结构越复杂，配合物的稳定常数就越大，这些是金属离子方面影响配合物稳定性大小的本质因素。此外，溶液的酸度、温度和其他配体的存在等外界条件的变化也影响配合物的稳定性。

2. Side Reaction and Its Coefficients

2. 副反应及副反应系数

In the actual analysis work, the complexometric titration is performed under certain conditions. For example, in order to control the acidity of the solution, it is necessary to add a certain buffer solution; or in order to mask the interfering ions, it is necessary to add a certain masking agent. In the complexometric titration under such condition, besides the main reactions of M and Y, the following side reactions may also occur:

实际分析工作中，配位滴定在一定的条件下进行的。例如，为控制溶液的酸度，需要加入某种缓冲溶液；为掩蔽干扰离子，需要加入某种掩蔽剂等。在这种条件下配位滴定，除了 M 和 Y 的主反应外，还可能发生如下一些副反应：

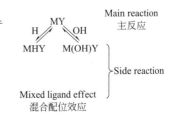

In the formula: L is the auxiliary ligand; N is the interfering ion.

The side reaction of reactant M or Y is not conducive to the main reaction. The side reaction of MY is beneficial to the main reaction, but most of these mixed complexes are not stable and can be neglected. The following part mainly discusses the acid effect and the ligand effect that have a greater impact on the coordination equilibrium.

(1) The acid effect of EDTA and the coefficient K_{MY} in the formula (4.2) describes at what degree of complexing reaction is going without any side reactions. When Y and H undergo side reactions, in addition to free dissociative Y, the ligands that are not coordinated with metal ions include HY, H_2Y, ... , H_6Y, etc. Therefore, the concentration of EDTA that is not complexed with M should be equal to that of the above seven species, which is shown as Y′:

$$Y' = [Y] + [HY] + \cdots + [H_6Y] \tag{4.3}$$

Due to the side reaction between hydrogen ion and Y, the ability of EDTA to participate in the main reaction is reduced. This phenomenon is called **acid effect**. The degree of its influence can be weighed by the acid effect coefficient $\alpha_{Y(H)}$:

$$\alpha_{Y(H)} = \frac{[Y']}{[Y]} \tag{4.4}$$

$\alpha_{Y(H)}$ indicates the ratio of the total concentration of EDTA that is not complexed with metal ions to the concentration of the free dissociative Y at a certain pH. Obviously, $\alpha_{Y(H)}$ is the reciprocal of Y's distribution fraction δ_Y. That is:

$$\alpha_{Y(H)} = \frac{[Y]+[HY]+\cdots+[H_6Y]}{[Y]} = \frac{1}{\delta_Y}$$

After deduction, we can get:

$$\alpha_{Y(H)} = 1 + \frac{[H^+]}{K_{a6}} + \frac{[H^+]^2}{K_{a5}K_{a6}} + \cdots + \frac{[H^+]^6}{K_{a1}K_{a2}K_{a3}K_{a4}K_{a5}K_{a6}} \tag{4.5}$$

In the formula, K_{a1}, K_{a2}, ... , K_{a6} are all the dissociation constants of EDTA. Based on them, the $\alpha_{Y(H)}$ values at any pH can be calculated according to the formula (4.5). When $\alpha_{Y(H)}=1$, it indicates that there is no side reaction. The bigger the $\alpha_{Y(H)}$ is, the greater the acid effect has.

Example 4-1

Calculate pH the $\alpha_{Y(H)}$ of EDTA when pH=5.0.

Solution

The dissociation constants $K_{a1} \sim K_{a6}$ of EDTA are known: $10^{-0.9}, 10^{-1.6}, 10^{-2.0}, 10^{-2.67}, 10^{-6.16}, 10^{-10.26}$. So at pH=5, we can get:

$$\alpha_{Y(H)} = 1 + \frac{[H^+]}{K_{a6}} + \frac{[H^+]^2}{K_{a5}K_{a6}} + \cdots + \frac{[H^+]^6}{K_{a1}K_{a2}K_{a3}K_{a4}K_{a5}K_{a6}}$$

$$= 1 + \frac{10^{-5.0}}{10^{-10.26}} + \frac{10^{-10.0}}{10^{-16.42}} + \frac{10^{-15.0}}{10^{-19.09}} + \cdots + \frac{10^{-30}}{10^{-23.59}}$$

$$= 10^{6.45}$$

As a result, we can get: $\lg\alpha_{Y(H)} = 6.45$

The $\lg\alpha_{Y(H)}$ values at different pH values are listed in table 4-3. It can be seen that in most cases $\lg\alpha_{Y(H)}$ is not equal to 1, and [Y′] is always greater than [Y]. Only when pH>12, $\alpha_{Y(H)}$ then equals to 1, and EDTA is almost completely dissociated into Y; at such time, EDTA has the strongest coordination ability.

(2) Coordination effect of metal ions and its coefficient The **coordination effect** of metal ions refers to the side reaction caused by the coordination of other ligands in the solution (auxiliary ligands, ligands in buffer solutions or masking agents, etc.) with metal ions, which reduces the ability of metal ions to participate in the main reaction.

从表 4-3 可以看出，多数情况下 $\alpha_{Y(H)}$ 不等于 1，[Y′] 总是大于 [Y]，只有在 pH>12 时，$\alpha_{Y(H)}$ 才等于 1，EDTA 几乎完全解离为 Y，此时 EDTA 的配位能力最强。

（2）金属离子的配位效应及配位效应系数 金属离子的配位效应是指溶液中其他配体（辅助配体、缓冲溶液中的配体或掩蔽剂等）能与金属离子配位所产生的副反应，使金属离子参加主反应能力降低的现象。当有配位

Table 4-3 $\lg\alpha_{Y(H)}$ at different pH values
表 4-3 不同 pH 值的 $\lg\alpha_{Y(H)}$

pH	$\lg\alpha_{Y(H)}$	pH	$\lg\alpha_{Y(H)}$	pH	$\lg\alpha_{Y(H)}$
0.0	23.64	3.4	9.70	6.8	3.55
0.4	21.32	3.8	8.85	7.0	3.32
0.8	19.08	4.0	8.44	7.5	2.78
1.0	18.01	4.4	7.64	8.0	2.27
1.4	16.02	4.8	6.84	8.5	1.77
1.8	14.27	5.0	6.45	9.0	1.28
2.0	13.51	5.4	5.69	9.5	0.83
2.4	12.19	5.8	4.98	10.0	0.45
2.8	11.09	6.0	4.65	11.0	0.07
3.0	10.60	6.4	4.06	12.0	0.01

When there is a ligand effect, the metal ions that are not complexed to Y, in addition to the free dissociative M, also include ML, ML$_2$, ... , ML$_n$, etc. If [M′] represents the total concentration of metal ions un-complexed with Y, then ML goes:

效应存在时，未与 Y 配位的金属离子，除游离的 M 外，还有 ML、ML$_2$、…、ML$_n$ 等，以 [M′] 表示未与 Y 配位的金属离子总浓度。

$$[M']=[M]+[ML]+[ML_2]+[ML_3]+\cdots+[ML_n] \quad (4.6)$$

Since the coordination of L and M reduces [M] and affects the main reaction of M and Y, its influence can be expressed by $\alpha_{M(L)}$, the ligand effect coefficient will be:

由于 L 与 M 配位使 [M] 降低，影响 M 与 Y 的主反应，其影响可用配位效应系数 $\alpha_{M(L)}$ 表示。$\alpha_{M(L)}$ 表示如下：

$$\alpha_{M(L)} = \frac{[M']}{[M]} = \frac{[M]+[ML]+[ML_2]+[ML_3]+\cdots+[ML_n]}{[M]} \quad (4.7)$$

$\alpha_{M(L)}$ indicates the ratio of the total concentration of various species of metal ions that are not complexed with Y to the concentration of free dissociative metal ions. When $\alpha_{M(L)}$=1, [M′]=[M] means that the metal ion is free from side reactions. The bigger $\alpha_{M(L)}$ is, the greater the side reaction has.

未与 Y 配位的金属离子的各种型体的总浓度是游离金属离子浓度的多少倍。当 $\alpha_{M(L)}$=1 时，[M′]=[M] 表示金属离子没有发生副反应，$\alpha_{M(L)}$ 值越大，副反应就越严重。

If $K_1, K_3, \ldots, K_n$ are used to represent the stability constants of the complex ML$_n$ at all levels. K's equation is taken into the formula (4.7) and after sorting out, we can get:

若用 K_1、K_3、…、K_n 表示配合物 ML$_n$ 的各级稳定常数，可将 K 的关系式代入式（4.7），并整理得式（4.8）。

coordination equilibrium

stability constants at all levels

$$M+L \rightleftharpoons ML \qquad K_1 = \frac{[ML]}{[M][L]}$$

$$ML+L \rightleftharpoons ML_2 \qquad K_2 = \frac{[ML_2]}{[ML][L]}$$

$$\vdots \qquad \vdots$$

$$ML_{n-1}+L \rightleftharpoons ML_n \qquad K_n = \frac{[ML_n]}{[ML_{n-1}][L]}$$

$$\alpha_{M(L)}=1+[L]K_1+[L]^2K_1K_2+\cdots+[L]^nK_1K_2\cdots K_n \quad (4.8)$$

Chemistry manuals often provide data on the cumulative stability constant (β_i) of the complex. The relation between β_i and stability constant K is:

化学手册中还常常给出配合物的累积稳定常数（β_i）的数据，β_i 与稳定常数 K 之间存在一定的关系。

$$\beta_1=K_1$$
$$\beta_2=K_1K_2$$
$$\ldots\ldots$$
$$\beta_n=K_1K_2\ldots K_n$$

β_i's equation is taken into the formula (4.8) and we can get:

$$\alpha_{M(L)} = 1 + \beta_1[L] + \beta_2[L]^2 + \cdots + \beta_n[L]^n \tag{4.9}$$

It can be seen that the greater the concentration of free ligand is or the bigger the stability constant of the complex is, the bigger the coordination effect coefficient will be, which is not conducive to the progress of the main reaction.

可以看出，游离配体的浓度越大，或其配合物稳定常数越大，则配位效应系数越大，不利于主反应的进行。

3. Conditional Stability Constant

3. 条件稳定常数

In the absence of any side reactions, the stability constant of the complex MY is signified by K_{MY}, which is free from external conditions such as solution concentration and acidity, so it is also called the **absolute stability constant**. When the coordination reaction of M and Y is made under certain acidity conditions, and there are other ligands besides EDTA, it will cause side reactions and affect the progress of the main reaction. At such time, the stability constant K_{MY} can no longer objectively reflect the progress degree of the main reaction. In the expression of the stability constant, Y should be replaced by Y′, M should be replaced by M′, and the stability constant of the complex should be expressed as:

在没有任何副反应存在时，配合物 MY 的稳定常数用 K_{MY} 表示，它不受溶液浓度、酸度等外界条件影响，所以又称绝对稳定常数。当 M 和 Y 的配合反应在一定的酸度条件下进行，并有 EDTA 以外的其他配体存在时，将会引起副反应，从而影响主反应的进行。此时，稳定常数 K_{MY} 已不能客观地反映主反应进行的程度。稳定常数的表达式中，Y 应以 Y′ 替换，M 应以 M′ 替换，这时配合物的稳定常数应表示为式（4.10）。

$$K'_{MY} = \frac{[MY]}{[Y'][M']} \tag{4.10}$$

The actual stability constant obtained by considering the influence of the side reaction becomes the **conditional stability constant** K'_{MY} and is a general expression of the conditional stability constant. Sometimes, in order to clearly indicate which component has a side reaction, we can put a sign "′" on the upper right of the symbol of the component where the side reaction occurs. In the complexometric titration, generally speaking, the side reactions that have a greater impact on the main reaction are the acid effect of EDTA and the ligand effect of metal ions, especially the acid effect. If other side reactions are left alone and only the acid effect of EDTA is considered, then the formula (4.10) becomes:

这种考虑副反应影响而得出的实际稳定常数称为条件稳定常数，K'_{MY} 是条件稳定常数的笼统表示，有时为明确表示哪个组分发生了副反应，可将 "′" 写在发生副反应的该组分符号的右上方。配位滴定法中，一般情况下，对主反应影响较大的副反应是 EDTA 的酸效应和金属离子的配位效应，其中尤以酸效应影响更大。如不考虑其他副反应，仅考虑 EDTA 的酸效应，则式（4.10）变为式（4.11）。

$$K'_{MY} = \frac{[MY]}{[Y'][M']} = \frac{K_{MY}}{\alpha_{Y(H)}} \quad (4.11)$$

The formula (4.11) is an important equation for discussing coordination equilibrium, which shows that the conditional stability constant of MY changes with the acidity of the solution.

式（4.11）其是讨论配位平衡的重要公式，表明 MY 的条件稳定常数随溶液的酸度而变化。

 Example 4-2

Assuming that only the acid effect is considered, please calculate the $K_{ZnY'}$ of ZnY at pH=2.0. and pH=5.0.

Solution

(1) at pH=2.0, after consulting the tabs, we find $\lg\alpha_{Y(H)}$=13.51, $\lg K_{ZnY'}$=16.50. So, we can get:

$\lg K_{ZnY'}$=16.50−13.51=2.99

$K_{ZnY'}=10^{2.99}$

(2) at pH=5.0, after consulting the tabs, we find $\lg\alpha_{Y(H)}$=6.45. Thus, we can get:

$\lg K_{ZnY'}$=16.50−6.45=10.05

$K_{ZnY'}=10^{10.05}$

The above calculations show that ZnY is stable at pH=5.0, while ZnY is unstable at pH=2.0. Therefore, in order to make the complexometric titration run smoothly and obtain accurate analysis and measurement results, appropriate acidity conditions must be selected.

以上计算表明，pH=5.0 时 ZnY 稳定，而 pH=2.0 时 ZnY 不稳定，所以为使配位滴定顺利进行，得到准确的分析测定结果，必须选择适当的酸度条件。

Section 3　Principle of Complexometric Titration
第 3 节　配位滴定法原理

Using the analysis of Ca^{2+} in water by its titration with EDTA as an example. The dashed lines shown in the titration curve (fig.4-3) indicate the volume of the EDTA titrant that is needed to reach the equivalence point and gives the Ca^{2+} concentration at this point by using the function pCa, where pCa=−log[Ca^{2+}]. The titration curve in this example shows the response expected for a 10.00mL aliquot of a 0.01000M Ca^{2+} solution in water at pH 10.00 that is titrated using 0.0050M EDTA.

以 EDTA 滴定水中钙离子为例，图 4-3 中虚线表示达到化学计量点所消耗的 EDTA 滴定剂的体积及此时溶液中 Ca^{2+} 的浓度 pCa，其中 pCa=−lg[Ca^{2+}]。本滴定曲线为在 pH 值为 10.00 时，用浓度为 0.0050mol·L⁻¹ EDTA 滴定 10.00mL 浓度为 0.01000mol·L⁻¹ Ca^{2+} 溶液的响应结果。

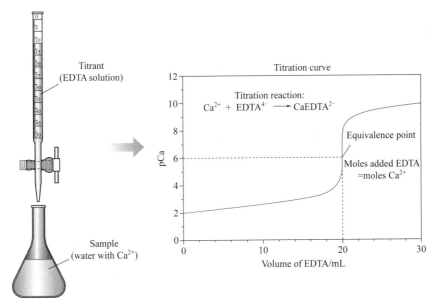

Fig.4-3 A typical complexometric titration

图 4-3　配位滴定装置及典型的滴定曲线

1. Titration Curve

Similar to the situation of acid-base titration, during the complexometric titration, in the solution of metal ions, the coordination reaction of metal ions occurs constantly with the addition of the coordination titrant while the concentration decreases as well. Near the stoichiometric point, the concentration of metal ions in the solution breaks. Figure 4-4 shows the titration curve of Ca^{2+} titrated by EDTA. Since Ca^{2+} is neither easy to be hydrolyzed nor to react with other complexing agents, only the acid effect of EDTA needs considering. The concentration of titrated Ca^{2+} in the solution at different stages can be calculated by the formula (4.11). The way is much similar to that of acid-base titration (figure 4-5). It can be seen from figure 4-4 that when EDTA is used to titrate Ca^{2+}, the curve before the equivalence point only follows the addition of EDTA and the concentration of Ca^{2+} keeps decreasing. The curve after it is affected by the acid effect of EDTA and the pCa varies with pH.

If the metal ion to be titrated is easily coordinated with other ligands or easily hydrolyzed, the titration curve is affected by both the acid effect and the ligand effect. Figure 4-6 is the titration curve of Ni^{2+} titrated by

1. 滴定曲线

与酸碱滴定情况相似，配位滴定时，在金属离子的溶液中，随着配位滴定剂的加入，金属离子不断发生配位反应，它的浓度也随之减小。在化学计量点附近，溶液中金属离子浓度发生突跃。图 4-4 为 EDTA 滴定 Ca^{2+} 的滴定曲线。由于 Ca^{2+} 既不易水解也不与其他配位剂反应，只需考虑 EDTA 的酸效应，利用式（4.11）即可计算不同阶段溶液中被滴定的 Ca^{2+} 的浓度，计算的思路类同于酸碱滴定，见图 4-5。从图 4-4 可以看出，用 EDTA 滴定 Ca^{2+}，在滴定终点前一段曲线的位置仅随 EDTA 的滴入，Ca^{2+} 的浓度不断减小，后一段受 EDTA 的酸效应影响，pCa 数值随 pH 不同而不同。

如果被滴定的金属离子是易与其他配体配合或易水解的离子，则滴定曲线同时受酸效应和配位效应影响。图 4-6 是 EDTA 滴定 Ni^{2+} 的滴定曲线，

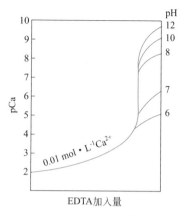

Fig. 4-4　The titration curve of 0.1 mol·L^{-1} Ca^{2+} titrated by 0.01 mol·L^{-1} EDTA

图 4-4　用 0.01 mol·L^{-1} EDTA 滴定 0.1 mol·L^{-1} Ca^{2+} 的滴定曲线

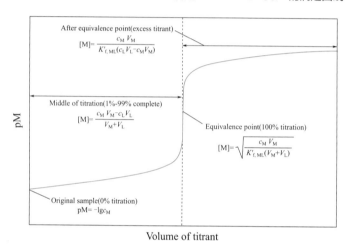

Fig.4-5　Equations for predicting the response for the titration of a metal ion(M) with a complexing agent(L) to form a 1∶1 complex

图 4-5　金属离子（M）与配位剂（L）形成 1∶1 配合物时滴定反应计算方程式

EDTA. Since Ni^{2+} is easily coordinated with NH$_3$ in the ammonia buffer solution, a relatively stable Ni(NH$_3$)$_4^{2+}$ is produced, which reduces the concentration of free dissociative Ni^{2+}, so the titration curve rises in the position before the stoichiometric point. The curve after the stoichiometric point is mainly affected by the acid effect of EDTA, as same as figure 4-4.

In the complexometric titration, the magnitude of the titration jump rests with the conditional stability constant of the complex K'_{MY} and the initial concentration of the metal ion. The bigger the conditional stability constant of the complex is, the larger the range of the titration jump has; when K'_{MY} is constant, the higher the initial concentration of metal ions is, the larger the range of the titration jump has.

由于在氨缓冲溶液中 Ni^{2+} 易与 NH$_3$ 配位，生成较稳定的 Ni(NH$_3$)$_4^{2+}$，使游离的 Ni^{2+} 的浓度减小，因而滴定曲线在化学计量点前一段的位置升高。化学计量点后段曲线的位置，主要受 EDTA 酸效应的影响，和图 4-4 的情况一样。

配位滴定中，滴定突跃的大小取决于配合物的条件稳定常数 K'_{MY} 和金属离子的起始浓度。配合物的条件稳定常数越大，滴定突跃的范围就越大；当 K'_{MY} 一定时，金属离子的起始浓度越大，滴定突跃的范围就越大。

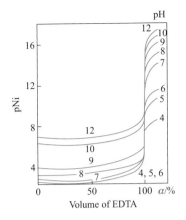

Fig. 4-6 The titration curve of 0.1 mol·L^{-1} Ni^{2+} titrated by 0.01 mol·L^{-1} EDTA ([NH$_3$]+[NH$_4^+$]=0.1 mol·L^{-1})

图 4-6 用 0.01 mol·L^{-1} EDTA 滴定 0.1 mol·L^{-1} Ni^{2+} 的滴定曲线（([NH$_3$]+[NH$_4^+$]=0.1 mol·L^{-1})）

2. Acid Effect Curve and Minimum pH for Titration of Metal Ions

From the example 4-2 in the previous section, we can see that at pH=2.0, the conditional stability constant K'_{ZnY} of ZnY is only $10^{2.99}$, which means an uncompleted coordination reaction. Obviously, titration cannot be performed under this acidity condition; When the acidity is decreased (that is, the pH is increased), lg$\alpha_{Y(H)}$ becomes smaller, which is beneficial to the formation of more complexes, which means a more completed coordination reaction. At pH=5.0, $K'_{ZnY}=10^{10.05}$, and it is shown that ZnY is quite stable and can be analyzed by titration. This indicates that for the complex ZnY, at pH=2.0-5.0, there is a line between titration and non-titration. Therefore, it is necessary to determine the maximum acidity allowed for the titration of different metal ions, that is, the minimum pH. In the complexometric titration, when the difference ΔpM between the visual end point and the stoichiometric point pM (pM=−lg[M]) is ±0.2 pM unit, and the allowable end point error is ±0.1%, according to the relevant formula, It can be deduced that the conditions for accurately determining a single metal ion are:

2. 酸效应曲线和滴定金属离子的最小 pH

从上节的例 4-2 中可以看到，在 pH=2.0 时，ZnY 的条件稳定常数 K'_{ZnY} 仅为 $10^{2.99}$，配位反应不完全，显然在该酸度条件下不能进行滴定；当将酸度降低（即提高 pH）时，lg$\alpha_{Y(H)}$ 变小，有利于形成更多的配合物，配合反应趋向完全，于 pH=5.0 时，$K'_{ZnY}=10^{10.05}$，说明 ZnY 已相当稳定，能够进行滴定分析。这表明，对于配合物 ZnY 来说，在 pH=2.0～5.0 之间，存在着可以滴定与不可以滴定的界限。因此，需要求出对不同的金属离子进行滴定时，允许的最高酸度，即最小 pH。在配位滴定中，当目测终点与化学计量点二者 pM(pM=−lg[M])）的差值 ΔpM 为 ±0.2pM 单位，允许的终点误差为 ±0.1% 时，根据有关公式，可推导出准确测定单一金属离子的条件。

$$\lg(cK'_{MY}) \geqslant 6 \tag{4.12}$$

In the formula: c refers to the concentration of metal ions.

For 10^{-2} mol·L^{-1} Zn^{2+}, the formula (4.12) can be rewritten as:

$$\lg K'_{ZnY} \geqslant 8$$

$\lg K_{ZnY}=16.50$ and $\lg K'_{ZnY} \geqslant 8$ are taken into the equation (4.11), we can get $\lg \alpha_{Y(H)} \leqslant 8.50$. After consulting table 4-3, when the pH is $\geqslant 4.0$, $\lg \alpha_{Y(H)} \leqslant 8.50$, and then $\lg K'_{ZnY} \geqslant 8$ can be ensured. Satisfying the requirement of $\lg(cK'_{MY}) \geqslant 6$, that is, for 10^{-2} mol·L^{-1} Zn^{2+}, at pH $\geqslant 4.0$, titration can be done; however, if pH<4.0, and accurate determination cannot be guaranteed, titration cannot be done. At pH=4.0, it is the minimum pH of 10^{-2} mol·L^{-1} Zn^{2+} for titration.

For different metal ions, the allowable minimum pH can be calculated. Figure 4-7 shows the minimum pH of 10^{-2} mol·L^{-1} metal ions when the allowable end point error is ±0.1%. The connected curve is called the **EDTA acid effect curve** from which we can easily find the minimum pH allowed for the titration of various metal ions. For example, $\lg K_{FeY} \geqslant 25.1$, we can find the pH=1.0. It is required when 10^{-2} mol·L^{-1}, Fe^{3+} is titrated, pH should be $\geqslant$ 1.0.

对于 10^{-2} mol·L^{-1} Zn^{2+}，式（4.12）变为：

对于 10^{-2} mol·L^{-1} 的 Zn^{2+}，将 $\lg K_{ZnY}$=16.50、$\lg K'_{ZnY} \geqslant 8$ 代入式（4.11）可得 $\lg \alpha_{Y(H)} \leqslant 8.50$，查表 4-3 可知，当 pH $\geqslant 4.0$ 时，就可使 $\lg \alpha_{Y(H)} \leqslant 8.50$。进而保证 $\lg K'_{ZnY} \geqslant 8$，满足 $\lg(cK'_{MY}) \geqslant 6$ 的要求，即对 10^{-2} mol·L^{-1} 的 Zn^{2+} 而言，当 pH $\geqslant 4.0$ 时，可以进行滴定；而 pH ＜ 4.0，就不能保证准确测定，因而不能滴定，pH=4.0 即为滴定 10^{-2} mol·l^{-1} 的 Zn^{2+} 的最小 pH。

对于不同的金属离子，可求出其允许的最小 pH。图 4-7 为 10^{-2} mol·L^{-1} 的金属离子在允许终点误差为 ±0.1% 时的最小 pH 所连成的曲线，称为 **EDTA 酸效应曲线**。从酸效应曲线可以方便地查到滴定各种金属离子允许的最小 pH。例如，$\lg K_{FeY} \geqslant 25.1$，可查得 pH=1.0，要求在滴定 10^{-2} mol·L^{-1} 的 Fe^{3+} 时，应使 pH $\geqslant 1.0$。

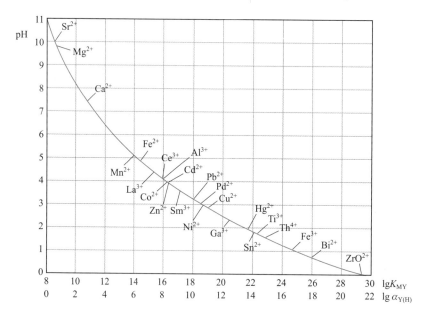

Fig.4-7 Acid Effect Curve(The concentration of metal ions is 0.01 mol·L^{-1}, and the relative error allowed is ±0.1%)

图 4-7 酸效应曲线（金属离子浓度为 0.01mol·L^{-1}，允许测定的相对误差为 ±0.1%）

When a certain metal ion is actually measured, the pH should be controlled within the range that is greater than the minimum pH and the metal ion does not undergo hydrolysis. Finally, it should be emphasized that the acid effect curve is derived under certain conditions and requirements, which only take the influence of acidity on EDTA into consideration, but neither thinking over the influence of acidity on metal ions and MY, nor the influence of other ligands. Therefore, it is rough and can only be used for reference. In actual analysis, the selection of appropriate acidity should be determined in conjunction with experiments.

实际测定某金属离子时,应将 pH 控制在大于最小 pH 且金属离子又不发生水解的范围之内。最后需强调指出,酸效应曲线是在一定条件和要求下得出的,只考虑了酸度对 EDTA 的影响,没有考虑酸度对金属离子和 MY 的影响,更没有考虑其他配体存在的影响,因此它是较粗糙的,只能提供参考。实际分析中,合适的酸度选择应结合实验来确定。

Section 4　Metallochromic Indicators
第 4 节　金属指示剂

1. Principle of Metallochromic Indicators

Metallochromic indicators are widely used in complexometric titration to indicate the end point. Some organic complexing agents of the metallochromic indicators can form colored complexes with the metal ion M, whose color is different from the color of the free indicator itself, thereby they are used for indicating the end of the titration. Now let's take Eriochrome black T (signified by In) as an example to illustrate the working principle of the metallochromic indicator. Eriochrome black T can form a relatively stable red complex with metal ions (Ca^{2+}, Mg^{2+}, Zn^{2+} and others). When pH=8～11, Eriochrome black T (EBT) itself presents blue color.

1. 金属指示剂的作用原理

在配位滴定中广泛采用金属指示剂来指示滴定终点。金属指示剂是一些有机配位剂,能同金属离子 M 形成有色配合物,其颜色与游离指示剂本身颜色不同,从而指示滴定的终点。现以铬黑 T(以 In 表示)为例,说明金属指示剂的作用原理。铬黑 T 能与金属离子(Ca^{2+}、Mg^{2+}、Zn^{2+} 等)形成比较稳定的红色配合物,当 pH=8～11 时,铬黑 T 本身呈蓝色。

Metallochromic indicators
金属指示剂变色演示

During the determination, a small amount of EBT is added into the solution which contains the above-mentioned

测定时,在含上述金属离子的溶液中加入少量铬黑 T,这时有少量 MIn

$$In + M \rightleftharpoons MIn$$
（Blue）　　（Red）

metal ions. Then a small amount of MIn is generated and the solution appears the color of red. With the addition of EDTA, the free dissociative metal ions are gradually combined by EDTA to form MY. EDTA doesn't keep dripping until most of the free metal ions are coordinated, and at such time because the conditional stability constant of complex MY is larger than that of complex MIn, a slight excess of EDTA will deprive the M from MIn and make the indicator free. The red solution suddenly turns to blue, indicating the coming of the end point.

$$\text{MIn} + \text{Y} \rightleftharpoons \text{MY} + \text{In}$$
$$\text{(Red)} \qquad\qquad \text{(Blue)}$$

Many metallochromic indicators not only have the properties of ligands, but at different pH values, the indicators themselves will show different colors. For example, EBT is a kind of triprotic weak acid, which can present different colors with the changes of pH of the solution. When pH<6, EBT appears red; when pH>12, it also appears orange. Obviously, when pH<6 or pH>12, there is no significant difference between the color of free EBT and the color of complex MIn. Only when titration is performed under the acidity conditions of pH 8～11, the color converts from red to blue at the end point suddenly. Therefore, when choosing a metallochromic indicator, attention must be paid to choose a suitable pH range.

2. Qualifications for Metallochromic Indicators

As can be seen from the above example of EBT, the metallochromic indicators must meet the following requirements:

① At the pH of titration, the color of the free indicator (In) itself should be significantly different from that of MIn coordinated by the indicator with metal ions;

② The color reaction of the colored complex formed by the metal ion and the indicator should be sensitive; that is to say, the color can still be expressed when the concentration of the metal ion is very small.

③ The metal ions and indicator complex MIn should maintain proper stability. On the one hand, it should be less than the stability of MY complex, that is, $K_{MIn} < K_{MY}$, so that when EDTA is titrated to the equivalence point, the indicator can be replaced from the MIn complex. But on the other hand, if the stability of MIn is too poor, before reaching the equivalence point, the color of the indicator itself will be shown, which means the end point will appear earlier, errors will be introduced and color changes will become less acute.

3. Possible Problems With Metallochromic Indicators

(1) Blocking of indicators   Some indicators can produce extremely stable complexes with certain metal ions. Compared with the corresponding MY complexes, these complexes are so stable that when the equivalence point is reached and excess EDTA is added, the indicator cannot be released and the color of the solution does not change. This is called **the blocking phenomenon of the indicator**. For example, EBT is used as an indicator. When Ca^{2+} and Mg^{2+} are titrated by EDTA at pH=10, Fe^{3+}, Al^{3+}, Ni^{2+} and Co^{2+} have a blocking effect on EBT. At this time, a small amount of triethanolamine (for masking Fe^{3+} and Al^{3+}) and KCN (for masking Ni^{2+} and Co^{2+}) can be added to eliminate interference.

(2) Ossification of indicators   Some indicators and metal ion complexes have very low solubility in water, which makes the replacement between EDTA and indicator metal ion complex MIn slow, and the color change at the end point not obvious. This phenomenon is called **ossification of indicators**. At such time, appropriate organic solvent can be added or rise of solubility by heating. For example, when PAN is used as the indicator, a small amount of methanol or ethanol can be added, or the solution can be heated appropriately to speed up the replacement speed and make the indicator's color change more sensitive.

(3) Oxidative deterioration of indicators   Most of the

③ 金属离子与指示剂配合物 MIn 应有适当的稳定性。一方面应小于配合物 MY 的稳定性，$K_{MIn} < K_{MY}$，这样才能使 EDTA 滴定到化学计量点时，将指示剂从 MIn 配合物中取代出来。但是另一方面，如果 MIn 的稳定性太差，则在到达化学计量点前，就会显示出指示剂本身的颜色，使终点提前出现，而引入误差，颜色变化也不敏锐。

3. 使用金属指示剂可能出现的问题

（1）指示剂的封闭　有的指示剂能与某些金属离子生产极稳定的配合物，这些配合物较对应的 MY 配合物更稳定，以致达到化学计量点时滴入过量 EDTA，指示剂也不能释放出来，溶液颜色不变化，这叫指示剂的封闭现象。例如以铬黑 T 作指示剂，在 pH=10 的条件下，用 EDTA 滴定 Ca^{2+}、Mg^{2+} 时，Fe^{3+}、Al^{3+}、Ni^{2+} 和 Co^{2+} 对铬黑 T 有封闭作用，这时，可加入少量三乙醇胺（以掩蔽 Fe^{3+} 和 Al^{3+}）和 KCN（掩蔽 Ni^{2+} 和 Co^{2+}）以消除干扰。

（2）指示剂的僵化　有些指示剂和金属离子配合物在水中溶解的溶解度很小，使 EDTA 与指示剂金属离子配合物 MIn 的置换缓慢，终点的颜色变化不明显，这种现象称为指示剂僵化，这时可加入适当的有机溶剂或加热，以增大其溶解度。例如，用 PAN 作指示剂时，可加入少量的甲醇或乙醇，也可将溶液适当加热以加快置换速度，使指示剂的变色敏锐一些。

Blocking of indicators
指示剂的封闭

Ossification of indicators
指示剂的僵化

metal indicators are colored compounds with many double bonds, which are easily decomposed by sunlight, oxidizers, and air. Some indicators are unstable in aqueous solutions and deteriorate over time. For example, the aqueous solutions of EBT and calconcarboxylic acid are easy to oxidize and deteriorate, so they are often prepared as a solid mixture or added with reducing substances to prepare a solution, such as adding hydroxylamine hydrochloride as a reducing agent.

4. Commonly Used Metallochromic Indicators

Commonly used metallochromic indicators are shown in table 4-4.

In addition to the indicators listed in table 4-4, there is also one called Cu-PAN, which is a mixed solution of Cu-EDTA and a small amount of PAN. Such indicator can titrate many metal ions, especially for some ions that are not stable enough to coordinate with PAN or whose colors are not sufficiently expressed. For example, at pH=10, this indicator is used to titrate Ca^{2+} with EDTA. The color changing process goes in the following way: Initially, the concentration of Ca^{2+} in the solution is higher, which can capture Y from CuY and then form the CaY. The free dissociative Ca^{2+} is coordinated with PAN to show a violet color. The reaction formula can be expressed as follows:

$$CuY + PAN + Ca^{2+} \rightleftharpoons CaY + Cu\text{-}PAN$$
$$\text{(Blue)}\quad\text{(Yellow)}\qquad\text{(Colorless)}\quad\text{(Violet)}$$

When titrating with EDTA, EDTA is first coordinated with the free Ca^{2+}, and finally the PAN in Cu-PAN is replaced by EDTA and becomes CuY and PAN. The green color is formed by the mixture of the two, that is to say, the end point is reached. The Cu-PAN indicator can be used within a wide pH range (pH=2~12) with Ni^{2+} having a blocking effect on it. In addition, when this indicator is put into use, a masking agent that can participate in Cu^{2+} to form a more stable complex cannot be used at the same time.

$$Cu\text{-}PAN + Y \longrightarrow CuY + PAN$$
$$\text{(Green)}$$

（3）指示剂的氧化变质　金属指示剂大多数是具有许多双键的有色化合物，容易被日光、氧化剂、空气所分解，有些指示剂在水溶液中不稳定，日久会变质。如铬黑T、钙指示剂的水溶液均易氧化变质，所以常配成固体混合物或加入具有还原性的物质来配成溶液，如加入盐酸羟胺等还原剂。

4. 常用的金属指示剂

常用的金属指示剂见表4-4。

除表4-4所列指示剂，还有一种Cu-PAN指示剂，它是Cu-EDTA与少量PAN的混合溶液。一些与PAN配位不够稳定的或不够显色的离子，可用此指示剂进行滴定。

例如，在pH=10时，用此指示剂，以EDTA滴定Ca^{2+}，其变色过程是：最初，溶液中Ca^{2+}浓度较高，它能夺取CuY中的Y，形成CaY，游离出来的Cu^{2+}与PAN配位而显紫红色。

用EDTA滴定时，EDTA先与游离的Ca^{2+}配位，最后Cu-PAN中的PAN被EDTA置换又成CuY及PAN，二者混合呈绿色，即达到终点。Cu-PAN指示剂可在很宽的pH范围（pH=2~12）内使用，Ni^{2+}对它有封闭作用。使用此指示剂时，不能同时使用能使Cu^{2+}形成更加稳定配合物的掩蔽剂。

Table 4-4 Frequently-used metallochromic indicators
表4-4 常用的金属指示剂

Indicators	pH ranges	Change of color		Direct titration of ions	Preparation of indicators	Attentions
		In	MIn			
Eriochrome black T 铬黑T，简称BT 或EBT	8～10	Blue	Red	pH=10, Mg^{2+}, Zn^{2+}, Cd^{2+}, Pb^{2+}, Mn^{2+}, Rare earth element ions(稀土元素离子)	1:100 NaCl (Solid)	Fe^{3+}, Al^{3+}, Cu^{2+}, Ni^{2+} and otherions block EBT
Acid chrome blue K 酸性铬蓝K	8～13	Blue	Red	pH=10, Mg^{2+}, Zn^{2+}, Mn^{2+} pH=13, Ca^{2+}	1:100 NaCl (Solid)	
Xylenol Orange 二甲酚橙，简称XO	<6	Bright Yellow	Red	pH<1, ZrO^{2+} pH=1～3.5, Bi^{3+}, Th^{4+} pH=5～6, Tl^{3+}, Zn^{2+} Pb^{2+}, Cd^{2+}, Hg^{2+}, Rare earth element ions	0.5% 水溶液 (5g·L^{-1})	Fe^{3+}, Al^{3+}, Ni^{2+}, Ti^{4+} and other ions block XO
Sulfo-salicylic acid 磺基水杨酸，简称Ssal	1.5～2.5	colorless	Violet	pH=1.5～2.5, Fe^{3+}	0.5% 水溶液 (50g·L^{-1})	Ssal is colorless, FeY^- shows yellow
Calcon-carboxylic acid 钙指示剂，简称NN	12～13	Blue	Red	pH=12～13, Ca^{2+}	1:100 NaCl (Solid)	Ti^{4+}、Fe^{3+}、Al^{3+}、Ni^{2+}、Cu^{2+}、Co^{2+}、Mn^{2+} and other ions block NN
[1-2-(pyridylazo)-2-naphthol] 1-(2-吡啶偶氮)-2-萘酚，简称PAN	2～12	Yellow	Violet	pH=2～3, Th^{4+}, Bi^{3+} pH=4～5, Cu^{2+}, Ni^{2+} Pb^{2+}, Cd^{2+}, Zn^{2+}, Mn^{2+}, Fe^{2+}	0.1% 乙醇溶液 (1g·L^{-1})	MIn has a low solubility in water. In order to prevent PAN from ossifying, it needs to be heated during titration

Guide to EDTA titrations of common metals is shown in fig.4-8.

图4-8为常见金属的EDTA滴定指南。

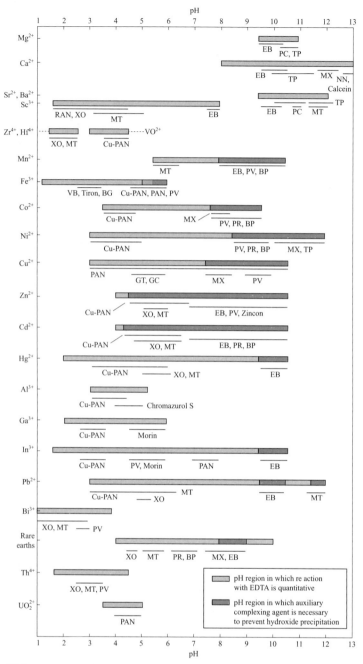

Fig.4-8 Guide to EDTA titrations of common metals (Light color shows pH region in which reaction with EDTA is quantitative. Dark color shows pH region in which auxiliary complexing agent is required to prevent metal from precipitating.)

Abbreviations for indicators:

BG, Bindschedler's green leuco base
GC, Glycinecresol red
MX, Murexide
Cu-PAN, PAN plus Cu-EDTA
PV, Pyrocatechol violet
XO, Xylenol orange

BP, Bromopyrogallol red
GT, Glycinethymol blue
NN, Patton & Reeder's dye
PC, o-Cresolphthalein complexone
TP, Thymolphthalein complexone

EB, Eriochrome black T
MT, Methylthymol blue
PAN, Pyridylazonaphthol
PR, Pyrogallol red
VB, Variamine blue B base

图 4-8 常见金属的 EDTA 滴定指南（浅色表示与 EDTA 定量反应的 pH 区域，深色表示需要辅助络合剂以防止金属沉淀的 pH 区域。）

Section 5 Methods to Improve the Selectivity of Complexometric Titration

第5节 提高配位滴定选择性的方法

Since EDTA can form stable complexes with most metal ions, there often exist multiple metal ions in the titrated test solution at the same time, which will interfere with the titration. How to improve the selectivity of coordination titration is a key issue to be solved in complexometric titration. In order to reduce or eliminate the interference of coexisting ions, the following methods are commonly used in actual titration.

由于 EDTA 能和大多数金属离子形成稳定的配合物，而在被滴定的试液中往往同时存在多种金属离子，这样将干扰滴定的进行。如何提高配位滴定的选择性是配位滴定要解决的重要问题。为了减少或消除共存离子的干扰，在实际滴定中，常用下列几种方法。

1. Control the Acidity of the Solution

1. 控制溶液的酸度

Since the stability constants of complexes formed by different metal ions and EDTA are not the same, the minimum pH allowed during titration is also different. If there are two or more metal ions existing in the solution at the same time, and the gap between stability constant of the complex formed by them and EDTA is also large enough, the acidity of the solution is controlled so that it only meets the minimum pH allowed by titration of a certain ion. But it will not cause the ion to hydrolyze and precipitate out. At this time, only one ion can form a stable complex with EDTA, and other ions will not undergo coordination reactions with EDTA; in this way, interference can be avoided.

不同的金属离子和 EDTA 所形成的配合物稳定常数是不相同的，因此在滴定时所允许的最小 pH 也不同。若溶液中同时有两种或两种以上的金属离子，它们与 EDTA 所形成的配合物稳定常数又相差足够大，则需控制溶液的酸度，使其只满足滴定某一种离子允许的最小 pH，但又不会使该离子发生水解而析出沉淀，此时就只能有一种离子与 EDTA 形成稳定的配合物，而其他离子与 EDTA 不发生配位反应，这样就可以避免干扰。

Assuming that there are two metal ions of M and N in the solution, both of them can form complexes with EDTA, but $K_{MY}>K_{NY}$, for the complexometric titration when interfering ions coexist, a relative error of $\leqslant \pm 0.5\%$ is usually allowable. When $c_M=c_N$ and when the indicator is used to detect the end point, the pM difference between the end point and the stoichiometric point, $\Delta pM \approx 0.3$. After calculation and deduction, it can be concluded that in order to accurately titrate M without N interference, it must satisfy:

设溶液中有 M 和 N 两种金属离子，它们均可与 EDTA 形成配合物，但 $K_{MY}>K_{NY}$，对于有干扰离子共存时的配位滴定，通常允许有 $\leqslant \pm 0.5\%$ 的相对误差。当 $c_M=c_N$ 而且用指示剂检测终点时终点与化学计量点二者 pM 的差值 $\Delta pM \approx 0.3$，经计算推导，可得出要准确滴定 M，而 N 不干扰，就要满足 $\Delta \lg K \geqslant 5$。

$$\Delta \lg K \geqslant 5 \tag{4.13}$$

In general, this formula is used as a condition for judging whether the acidity control can be used for separate titration.

For example, when the concentrations of Bi^{3+}, Pb^{2+} in the solution are both 10^{-2} mol·L^{-1}, we choose to titrate Bi^{3+}. From table 4-2 we can see that when $\lg K_{BiY}$=27.94, $\lg K_{PbY}$=18.04, $\Delta \lg K$=27.94−18.04=9.90, we can choose to titrate Bi^{3+} without Pb^{2+}'s interference. Furthermore, based on $\lg \alpha_{Y(H)} \leqslant \lg K_{MY}-8$, the minimum pH allowable for titration can be determined. In this case, $[Bi^{3+}]=10^{-2}$ mol·L^{-1}, from the acid effect curve of EDTA (figure 4-7), the minimum pH allowable for titration can be directly found and it is about 0.7; that is to say, titration of Bi^{3+} is done when pH $\geqslant$ 0.7. But the pH should not be too high during titration. When pH $\approx$ 2, Bi^{3+} will begin to hydrolyze and precipitate out, and Bi^{3+} is to be titrated in the solutions of Bi^{3+}, Pb^{2+}, the suitable acidity range is pH=0.7 ~ 2. At such time, Pb^{2+} does not coordinate with EDTA and does not interfere with the determination of Bi^{3+}.

Discussion

If the mixture contains Fe^{3+}, Al^{3+}, would it be possible just like the mixture with Bi^{3+}, Pb^{2+} to achieve their separate titration by acidity control?

TIPS

(1) First, discuss the possibility of separate titration based on their $\lg K_{MY}$. If they can be titrated separately, what about the pH? What can act as the indicator?

(2) What is the effect of Al^{3+} on the metal indicator? Can it be titrated directly? Attention: In the practice of coordination titration, we should not only consider the stability of the complex formed by EDTA and metal ions, but should take the interference of coexisting ions and the blocking of indicators into consideration as well. When these problems exist, different titration methods must be used to achieve coordinated titration.

2. Methods of Masking and Demasking

If the condition of $\Delta\lg(cK) \geqslant 5$ cannot be satisfied, N will be titrated at the same time during the titration of M and interference will occur. To overcome or eliminate this interference and further to improve the selectivity of titration, other measures must be taken, such as the method of masking, pre-separation method, or the application of other titrants. Commonly used masking methods are: coordination masking, precipitation masking and redox reaction masking, among which coordination masking is used frequently.

(1) Coordination masking This is a method of reducing the concentration of interfering ions to eliminate interference by making the use of coordination reaction. For example, when EDTA is used to titrate Ca^{2+}, Mg^{2+} in water to determine the hardness of water, the presence of Fe^{3+}, Al^{3+} will interfere with the determination. If triethanolamine is added to form a more stable complex with Fe^{3+}, Al^{3+}, then the interference can be removed. For another example, when Al^{3+} and Zn^{2+} coexist, NH_4F can be used to mask Al^{3+} to produce a more stable AlF_6^{3-} coordination ion. By adjusting pH=5-6, EDTA can be used to titrate Zn^{2+}, and Al^{3+} will not interfere.

It can be seen from the above examples that the coordination masking agent must meet the following requirements:

① The stability of the complex formed with the interfering ion must be greater than that of the complex formed by the EDTA with the ion, and what's more, these complexes should be colorless or light-colored, which does not affect the observation of the end point.

② The masking agent cannot form a complex with the measured ion, or the stability of the complex formed is much lower than that of the complex formed by the measured ion with EDTA, which will not affect the titration.

③ The masking agent not only should have a certain pH range, but it must conform to the measured pH range as well.

Some commonly used coordination masking agents and their application ranges are listed in table 4-5.

(2) Precipitation masking This is a method of reducing the concentration by using interfering ions and masking agent to form a precipitate. For example, adding NaOH into

定而发生干扰。要克服或消除这种干扰，提高滴定的选择性，必须采取其他措施，如采用掩蔽方法，预先分离的方法，或者改用其他滴定剂来达到这个目的。常用的掩蔽方法按反应类型不同，可分为配位掩蔽法、沉淀掩蔽法和氧化还原掩蔽法，其中以配位掩蔽法用得最多。

（1）配位掩蔽法 这是利用配位反应降低干扰离子浓度以消除干扰的方法。例如，用 EDTA 滴定水中的 Ca^{2+}、Mg^{2+}，测定水的硬度时，Fe^{3+}、Al^{3+} 等离子的存在会干扰测定，若加入三乙醇胺使与 Fe^{3+}、Al^{3+} 生成更稳定的配合物，则可消除 Fe^{3+}、Al^{3+} 的干扰。又如，在 Al^{3+} 与 Zn^{2+} 共存时，可用 NH_4F 掩蔽 Al^{3+}，使其生成稳定性较好的 AlF_6^{3-} 配离子，调节 pH=5～6，可用 EDTA 滴定 Zn^{2+}，而 Al^{3+} 不干扰。

由上例可以看出，配位掩蔽剂必须具备下列条件：

① 与干扰离子形成配合物的稳定性，必须大于 EDTA 与该离子形成配合物的稳定性，而且这些配合物应为无色或浅色，不影响终点的观察。

② 掩蔽剂不能与被测离子形成配合物，或形成配合物的稳定性要比被测离子与 EDTA 所形成配合物的稳定性小得多，这样才不会影响滴定进行。

③ 掩蔽剂的应用有一定的 pH 范围，而且要符合测定的 pH 范围。

常用的配位掩蔽剂及其使用范围见表 4-5。

（2）沉淀掩蔽法 这是利用干扰离子与掩蔽剂形成沉淀以降低其浓度的方法。例如，在 Ca^{2+}、Mg^{2+} 两种离

Table 4-5 Commonly-used masking agents
表 4-5 常用的掩蔽剂

Name	pH ranges	Masked ions	Notes
KCN	pH>8	Co^{2+}, Ni^{2+}, Cu^{2+}, Zn^{2+}, Hg^{2+}, Cd^{2+}, Ag^+, Tl^+ as well as platinum group elements	
NH$_4$F	pH=4～6	Al^{3+}, Ti^{4+}, Sn^{4+}, Zr^{4+}, W^{6+}, etc.	NH$_4$F is better than NaF, for the pH of the solution does not change much after adding it 用 NH$_4$F 比 NaF 好，优点是加入后溶液 pH 变化不大
	pH=10	Al^{3+}, Mg^{2+}, Ca^{2+}, Sr^{2+}, Ba^{2+} as well as rare earth elements	
TEA 三乙醇胺	pH=10	Al^{3+}, Sn^{4+}, Ti^{4+}, Fe^{3+}	to improve the masking effect with KCN 与 KCN 并用，可提高掩蔽效果
	pH=11～12	Fe^{3+}, Al^{3+} and small amount of Mn^{2+}	
Dimercaprol 二巯基丙醇	pH=10	Hg^{2+}, Cd^{2+}, Zn^{2+}, Bi^{3+}, Pb^{2+}, Ag^+, As^{3+}, Sn^{4+} and small amount of Co^{2+}, Ni^{2+}, Cu^{2+}, Fe^{3+}	
DDTC 铜试剂	pH=10	Work with Cu^{2+}, Hg^{2+}, Pb^{2+}, Cd^{2+}, Bi^{3+} to generate precipitation, where Cu-DDTC is brown, Bi-DDTC is yellow, so the present in an amount should be less than 2mg and 10mg respectively	
Tartaric acid 酒石酸	pH=1.2	Sb^{3+}, Sn^{4+}, Fe^{3+} and Cu^{2+} below 5mg	In the presence of ascorbic acid 在抗坏血酸存在下
	pH=2	Fe^{3+}, Sn^{4+}, Mn^{2+}	
	pH=5.5	Fe^{3+}, Al^{3+}, Sn^{4+}, Ca^{2+}	
	pH=6～7.5	Mg^{2+}, Cu^{2+}, Fe^{3+}, Al^{3+}, Mo^{4+}, Sb^{3+}, W^{6+}	
	pH=10	Al^{3+}, Sn^{4+}	

a solution in which two ions of Ca^{2+}, Mg^{2+} coexist and making pH>12. Then Mg^{2+} will generate *precipitation of* $Mg(OH)_2$, and Ca^{2+} is titrated by EDTA. In practice, the precipitation masking has certain limitations, for it requires that the formed precipitate be compact, have a small solubility, be colorless or light-colored, and have a small adsorption effect. Otherwise, dark color and large volume may lead to the adsorption of the ions to be measured or adsorption of indicator, which will affect the end point observation and measurement results. Some commonly used precipitation masking agents and their scope of application are listed in table 4-6.

(3) Redox reaction masking　　This is a method of using redox reactions to change the valence of interfering ions to eliminate interference. For example, when

子共存的溶液中加入 NaOH 溶液，使 pH>12，则 Mg^{2+} 生成 $Mg(OH)_2$ 沉淀，可以用 EDTA 滴定 Ca^{2+}。沉淀掩蔽法在实际应用中有一定的局限性，它要求所生成的沉淀致密，溶解度要小，无色或浅色，且吸附作用小。否则，颜色深，体积大，吸附待测离子或吸附指示剂都将影响终点的观察和测定结果。常用的一些沉淀掩蔽剂及其使用范围列于表 4-6 中。

（3）氧化还原掩蔽　这是利用氧化还原反应改变干扰离子价态以消除干扰的方法。例如，用 EDTA 滴定 Bi^{3+}、

Table 4-6 Precipitation masking agents in the complexometric titration

表 4-6 配位滴定中常用的沉淀掩蔽剂

Name	Masked ion	Ion to be measured	pH range	Indicators
NH_4F	Ca^{2+}, Sr^{2+}, Ba^{2+}, Mg^{2+}, Ti^{4+}, Al^{3+} and rare earth	Zn^{2+}, Cd^{2+}, Mn^{2+} (In the presence of reducing agent)	10	EBT
NH_4F	ditto	Cu^{2+}, Co^{2+}, Ni^{2+}	10	murexide
K_2CrO_4	Ba^{2+}	Sr^{2+}	10	Mg-EDTA EBT
Na_2S or DDTC	traces of heavy metals	Ca^{2+}, Mg^{2+}	10	EBT
H_2SO_4	Pb^{2+}	Bi^{3+}	1	xylenol orange
$K_4[Fe(CN)_6]$	traces of Zn^{2+}	Pb^{2+}	5-6	xylenol orange

EDTA is used to titrate Bi^{3+}, Zr^{4+}, Th^{4+}, if there exists Fe^{3+} in the solution, then Fe^{3+} *will* interferes with the determination. At such time, ascorbic acid or hydroxylamine hydrochloride can be added. Fe^{3+} will be reduced to Fe^{2+}, for the stability of the complex Fe^{2+} with EDTA is much lower than that of the complex with Fe^{3+} and EDTA ($\lg K_{FeY^-}=25.1$, $\lg K_{FeY^{2-}}=14.32$). Thus can mask the interference of Fe^{3+}. Commonly used reducing agents are: ascorbic acid, hydroxylamine hydrochloride, hydrazine, thiourea, cysteine, etc., some of which are also complexing agents.

(4) Demasking In the solution of metal ion complexes, add a reagent (demasking agent) to release the metal ions coordinated by EDTA or masking agent, and then to perform titration. This method is called **demasking**. For example, by the complexometric titration, Zn^{2+} and Pb^{2+} are determined in copper alloys. After the test solution displays alkaline property, add KCN to mask Cu^{2+} and Zn^{2+} (Warning: As a highly toxic substance, potassium cyanide is only allowed to be used in base solutions!). At this point, Pb^{2+} is not masked by KCN, so EBT can act as the indicator at pH=10. EDTA standard solution is used for titration and formaldehyde can be added to destroy $[Zn(CN)_4]^{2-}$ in the solution titrated with Pb^{2+}: Zn^{2+} which has been originally coordinated by CN^- is released again, and the titration is resumed with EDTA.

Zr^{4+}、Th^{4+} 等离子时，溶液中如果存在 Fe^{3+}，则 Fe^{3+} 干扰测定，此时可加入抗坏血酸或盐酸羟胺，将 Fe^{3+} 原为 Fe^{2+}，由于 Fe^{2+} 与 EDTA 配合物的稳定性比 Fe^{3+} 与 EDTA 配合物的稳定性小得多（$\lg K_{FeY^-}=25.1$，$\lg K_{FeY^{2-}}=14.32$），因而能掩蔽 Fe^{3+} 的干扰。常用的还原剂有抗坏血酸、盐酸羟胺、联胺、硫脲、半胱氨酸等，其中有些还原剂同时又是配位剂。

（4）解蔽方法 在金属离子配合物的溶液中，加入一种试剂（解蔽剂），将已被 EDTA 或掩蔽剂配位的金属离子释放出来，再进行滴定，这种方法叫解蔽。例如，用配位滴定法测定铜合金中的 Zn^{2+} 和 Pb^{2+}，试液调至碱性后，加 KCN 掩蔽 Cu^{2+}、Zn^{2+}（氰化钾是剧毒物，只允许在碱性溶液中使用！），此时 Pb^{2+} 不被 KCN 掩蔽，故可在 pH=10 时，以铬黑 T 为指示剂，用 EDTA 标准溶液进行滴定，在滴定 Pb^{2+} 后的溶液中，加入甲醛破坏 $[Zn(CN)_4]^{2-}$，原来被 CN^- 配位了的 Zn^{2+} 又可释放出来，再用 EDTA 继续滴定。

$$4HCHO + [Zn(CN)_4]^{2-} + 4H_2O \longrightarrow Zn^{2+} + 4H_2C\diagup^{OH}_{CN} + 4OH^-$$

In actual analysis, it is often not possible to obtain satisfactory results only with one masking agent. When many ions coexist, several masking agents or precipitating agents are always used in combination to obtain a better selectivity. But it should be noted that the amount of coexisting interference ions cannot be too much, otherwise satisfactory results will fail to be obtained.

3. Chemical Separation

When the methods like acidity control or masking have all failed to avoid interference, we still can turn to chemical separation which will separate the measured ions from other components. There are many separation methods.

4. Other Coordination Titrants

With the development of the complexometric titration, in addition to EDTA, some new amino-carboxy complexes have been developed as titrants. The stability of the complexes formed by these titrants with metal ions has their own characteristics and can be put into use to widen the choices in the complexometric titration. For example, the stabilities of the complexes formed by EDTA and Ca^{2+} and Mg^{2+} do not show great differences, while those by EGTA and Ca^{2+} and Mg^{2+} differ greatly. Thus, EGTA is used to selectively titrate Ca^{2+} when Ca^{2+} and Mg^{2+} coexist. The complex formed by EDTP and Cu^{2+} has high stability and can be used to selectively titrate Cu^{2+} in the solution with Cu^{2+}、Zn^{2+}、Cd^{2+} and Mn^{2+}.

在实际分析中，用一种掩蔽剂常不能得到令人满意的结果，当有许多离子共存时，常将几种掩蔽剂或沉淀剂联合使用，这样才能获得较好的选择性。但须注意共存干扰离子的量不能太多，否则得不到满意的结果。

3. 化学分离法

当利用控制酸度或掩蔽等方法避免干扰都有困难时，还可用化学分离法把被测离子从其他组分中分离出来，分离的方法很多。

4. 选用其他配位滴定剂

随着配位滴定法的发展，除EDTA外又研制了一些新型的氨羧配合物作为滴定剂，它们与金属离子形成配合物的稳定性各有特点，可以用来提高配位滴定法的选择性。例如，EDTA与Ca^{2+}、Mg^{2+}形成的配合物稳定性相差不大，而EGTA与Ca^{2+}、Mg^{2+}形成的配合物稳定性相差较大，故可以在Ca^{2+}、Mg^{2+}共存时，用EGTA选择性滴定Ca^{2+}。EDTP与Cu^{2+}形成的配合物稳定性高，可以在Cu^{2+}、Zn^{2+}、Cd^{2+}、Mn^{2+}共存的溶液中选择性滴定Cu^{2+}。

Section 6　Application of Complexometric Titration
第6节　配位滴定的应用

In the complexometric titration, various titration methods are used, for they can not only widen the application range of the complexometric titration, but also improve the selectivity of the complexometric titration.

在配位滴定中采用不同的滴定方式不但可以扩大配位滴定的应用范围，同时也可以提高配位滴定的选择性。

1. Titration Methods

(1) **Direct titration** As the most basic method in complexometric titration, direct titration is to turn the substance under test into a solution after preconditioning. After adjusting the acidity, adding an indicator, and sometimes adding an appropriate auxiliary ligand and masking agent, the EDTA standard solution is directly used for titration. Then according to the concentration of the standard solution and the volume consumed, calculate the content of the components to be tested in the test solution. Direct titration is applicable to:

At pH=1, to titrate Zr^{4+};
When pH=2～3, to titrate Fe^{3+}、Bi^{3+}、Th^{4+}、Ti^{4+}、Hg^{2+};
When pH=5～6, to titrate Zn^{2+}、Pb^{2+}、Cd^{2+}、Cu^{2+} and rare earth elements;
At pH=10, to titrate Mg^{2+}、Co^{2+}、Ni^{2+}、Zn^{2+}、Cd^{2+};
At pH=12, to titrate Ca^{2+}.

(2) **Back titration** When the ion under test is slowly coordinated with EDTA or hydrolyzed at the titrated pH, or has a blocking effect on the indicator, or there is no suitable indicator, the back titration can come into play. That is, first add a known amount of excess EDTA standard solution to coordinate with the measured ion, and then titrate the remaining EDTA with another standard solution of metal ions. The second titration's result shows how much of the excess reagent was used in the first titration, thus allowing the analyte's concentration to be calculated.

For example, it takes time for Al^{3+} to coordinate with EDTA; besides, Al^{3+} also has a blocking effect on indicators such as xylenol orange and is easier to hydrolyze; in this case, back titration is generally used. First add excess EDTA to the test solution, adjust the pH, heat and boil until the coordination of Al^{3+} with EDTA is done completely; After cooling, adjust pH=5-6, add xylenol orange, use Zn^{2+} standard solution to titrate the remaining EDTA.

(3) **Replacement titration** The replacement reaction is used to replace another metal ion or EDTA in the same

amount from the complex, and then perform titration. For example, when tin in tin bronze is measured, we can add excess EDTA to the test solution, then Sn^{4+} coordinates with EDTA along with coexisting Pb^{2+}、 Zn^{2+}、Cu^{2+}. Excess EDTA is removed by standard solution of Zn^{2+}. After adding NH_4F, the Y in SnY will be replaced by F^- and then Y can be titrated by the standard solution of Zn^{2+}, which means Sn can be calculated.

(4) Indirect titration Some metal ions such as Li^+, Na^+, K^+, Rb^+, Cs^+, etc. and some non-metal ions such as SO_4^{2-}, PO_4^{3-} etc. form unstable complexes with EDTA or cannot coordinate with EDTA. In such context, indirect titration can be used for determination. For example, the determination of Na^+ can be done when Na^+ is precipitated by uranyl zinc acetate and $[NaOAc \cdot Zn(OAc)_2 \cdot 3UO_3(OAc)_2 \cdot 9H_2O]$ precipitation is produced. After filtering, washing and dissolving the precipitate, it is quantified by titration of Zn^{2+} with EDTA. Another example is the determination of PO_4^{3-}. Under certain conditions, PO_4^{3-} can be precipitated as $MgNH_4PO_4$; then after filtering, dissolving the precipitate, and adjusting the solution at pH=10, EBT is used as an indicator, Mg^{2+} which has the same amount of PO_4^{3-} is titrated by EDTA. PO_4^{3-} can be calculated indirectly based on the amount of Mg^{2+}.

2. Application Examples

(1) Determination of the total hardness of water Industrial water often forms boiler scale which is mainly caused by calcium and magnesium carbonate, acid carbonate sulfate, chloride, etc. in the water. The content of calcium, magnesium salt, etc. in the water is represented by "hardness", where Ca^{2+} and Mg^{2+} are the main indicators for calculating hardness. The total hardness of water includes temporary hardness and permanent hardness. Calcium and magnesium salts existing in the form of carbonates and acid carbonates in water can be decomposed, precipitated and removed by heating. The hardness formed by these salts is called temporary hardness while the hardness formed by

物质的量的另一种金属离子或EDTA，然后进行滴定。例如测定锡青铜中的锡时，可向试液中加入过量的EDTA，Sn^{4+}与共存的Pb^{2+}、Zn^{2+}、Cu^{2+}等一起与EDTA配位，用Zn^{2+}标准溶液除去过量的EDTA，加NH_4F，F^-将SnY中的Y置换出来，再用Zn^{2+}标准溶液滴定置换出来的Y，即可求得Sn的含量。

（4）间接滴定法 有些金属离子，如Li^+、Na^+、K^+、Rb^+、Cs^+等和一些非金属离子如SO_4^{2-}、PO_4^{3-}等，由于和EDTA形成的配合物不稳定或不能与EDTA配位，这时可采用间接滴定的方法进行测定。例如，Na^+的测定可通过用醋酸铀酰锌来沉淀Na^+，生成醋酸铀酰锌$[NaOAc \cdot Zn(OAc)_2 \cdot 3UO_3(OAc)_2 \cdot 9H_2O]$沉淀，将沉淀过滤、洗涤、溶解后，以EDTA滴定$Zn^{2+}$而定量。又如$PO_4^{3-}$的测定，在一定条件下，可将$PO_4^{3-}$沉淀为$MgNH_4PO_4$，然后过滤、溶解沉淀，调节溶液的pH=10，以铬黑T作指示剂，以EDTA标准溶液滴定与PO_4^{3-}等物质的量的Mg^{2+}，由Mg^{2+}物质的量间接算出PO_4^{3-}的含量。

2. 配位滴定法应用示例

（1）水的总硬度测定 工业用水常形成锅垢，这是水中钙镁的碳酸盐、酸式碳酸盐、硫酸盐、氯化物等所致。水中钙、镁盐等的含量用"硬度"表示，其中Ca^{2+}、Mg^{2+}含量是计算硬度的主要指标。水的总硬度包括暂时硬度和永久硬度。在水中以碳酸盐及酸式碳酸盐形式存在的钙、镁盐，加热能被分解、析出沉淀而除去，这类盐所形成的硬度称为暂时硬度。而钙、镁的硫酸盐或氯化物等所形成的硬度称为永久硬度。

calcium and magnesium sulfate or chloride is called permanent hardness.

Hardness of water is an important indicator of industrial water. When boiler is supplied with water, hardness analysis is often performed to provide evidence for water treatment. To determine the total hardness of water is to determine the total content of Ca^{2+} and Mg^{2+} in the water. By complexometric titration, that is, in an ammonia buffer solution at pH=10, EBT is used as an indicator, and EDTA standard solution is used for direct titration until the solution turns from wine red to pure blue as the end point. During the titration, a small quantity of interfering ions such as Fe^{3+} and Al^{3+} in the water are masked with triethanolamine, Cu^{2+}, Pb^{2+}, Zn^{2+} and other heavy metal ions can be masked by KCN and Na_2S.

The total amount of calcium and magnesium ions in the measurement result is often calculated by the amount of calcium carbonate to determine the hardness of water. Different countries have different ways of expressing the hardness of water. Our country usually expresses the hardness by the mass concentration ρ containing $CaCO_3$, and the unit is $mg \cdot L^{-1}$. It is also expressed by the amount-of-substance concentration containing $CaCO_3$, and the unit is $mmol \cdot L^{-1}$. The national standard stipulates that the hardness of drinking water is calculated by $CaCO_3$ and cannot exceed 450 $mg \cdot L^{-1}$.

(2) Determination of aluminum hydroxide gel EDTA back titration is used to determine the aluminum content in aluminum hydroxide. First, dissolve a certain amount of aluminum hydroxide gel, and add $HAc-NH_4Ac$ buffer solution to control the acidity (pH=4.5). Then add an excess of EDTA standard solution, diphenylthizone as the indicator, and zinc standard solution is used to titrate until the solution turns from green-yellow to red, which marks the end point.

(3) Determination of ferric oxide, aluminum oxide, calcium oxide and magnesium oxide in silicate materials Silicates account for more than 75% of the earth's crust. Natural silicate minerals include quartz, mica, talc,

feldspar, dolomite, etc. while cement, glass, ceramic products, bricks, tiles, etc. are artificial ones. The main components of loess, clay, sand and other soils are also silicates. In addition to SiO_2, the composition of silicate mainly includes Fe_2O_3, Al_2O_3, CaO and MgO, etc. These components can usually be determined by EDTA complexometric titration. After the sample is pretreated into a test solution, at pH=2 ~ 2.5, sulfosalicylic acid as an indicator, EDTA standard solution is used to directly titrate Fe^{3+}. In the solution after titration of Fe^{3+}, add excess EDTA and adjust the pH to 4 ~ 5, PAN as an indicator, in the hot solution $CuSO_4$ standard solution is used to back titrate the excess EDTA to determine content of Al^{3+}. Take another test solution, add triethanolamine, KB as indicator at pH=10, and titrate the total amount of CaO and MgO with EDTA standard solution. Then take the same amount of test solution and add triethanolamine, adjust the pH> 12.5 with KOH solution, make Mg form precipitation of $Mg(OH)_2$, KB still as the indicator, EDTA standard solution is used to directly titrate to get the amount of CaO, and calculate the content of MgO by minusing method. This method is still widely used nowadays. The KB indicator used in the determination is a mixture of acid chrome blue K and naphthol green B.

Experiment 4   Preparation and Standardization of 0.05 mol·L^{-1} EDTA Standard Solution

1. Objective

(1) Master the preparation of EDTA standard solution by indirect method.
(2) Master the standardization method of EDTA standard solution.
(3) Understand the judgement of the end point by Eriochrome Black T (EBT) indicator.

Standardization of EDTA standard solution

2. Principle

Due to the low solubility of EDTA in water, $Na_2H_2Y_2 \cdot H_2O$ is usually used to prepare standard solution, which is also known as EDTA solution. $Na_2H_2Y_2 \cdot H_2O$ is a white, crystalline powder with molecular weight of 372.26. At room temperature, 11.1 g $Na_2H_2Y_2 \cdot H_2O$ can be dissolved in 100 mL water. EDTA standard solution can be prepared by indirect method, that is, appropriate amount of $Na_2H_2Y_2 \cdot H_2O$ is dissolved directly in water. The solution is then shaken well, and stored in a rigid glass bottle. The exact concentration of EDTA can be standardized using ZnO or Zn as primary standard and EBT or xylenol orange (XO) as indicator.

3. Apparatus and Materials

(1) Apparatus   buret (25mL), conical flask (250mL), volumetric flask (500 mL), beaker (50 mL).

(2) Materials   EDTA disodium salt ($Na_2H_2Y \cdot 2H_2O$), ZnO primary standard, dilute hydrochloric acid solution, methyl red indicator , ammonia solution, $NH_3 \cdot H_2O$-NH_4Cl buffer solution, Eriochrome Black T (EBT) indicator.

4. Procedures

(1) Preparation of 0.05 mol $\cdot$ L^{-1} EDTA standard solution   9.5g of $Na_2H_2Y_2 \cdot H_2O$ is weighed, and dissolved with water. The solution is then transferred to 500 mL volumetric flask , diluted to the mark with water, and shaked well. The prepared solution is transferred to a hard glass bottle for storage.

(2) Standardization of 0.05 mol $\cdot$ L^{-1} EDTA standard solution   Accurately weigh out approximately 0.12g of ZnO which is previously dried to constant weight at 800 ℃ . It is dissolved in 3mL of dilute hydrochloric acid, followed by adding 25 mL of distilled water and 1 drop of methyl red indicator. Ammonia solution is then added dropwise until the solution turns yellowish. After adding 25mL of distilled water,10mL of $NH_3 \cdot H_2O$-NH_4Cl buffer solution, and 2 drops of EBT indicator, the solution is titrated with EDTA solution until it changes from purplish red to pure blue.

The formula is as follows:

$$c_{EDTA} = \frac{w_{ZnO} \times 1000}{M_{ZnO} V_{EDTA}} \qquad M_{ZnO} = 81.38 \text{g} \cdot \text{mol}^{-1}$$

5. Notes

(1) The dissolution of $Na_2H_2Y \cdot 2H_2O$ in water is slow, so constantly stirring or heating is usually used to accelerate the dissolution.

(2) The prepared EDTA solution should be stored in a rigid glass bottle; otherwise, EDTA will react with the metal ions of the bottle, thereby affecting the concentration of EDTA.

(3) The titration process should not be too fast. In particular, the titrant should be added dropwise and constantly shaken when nearing end point of titration.

6. Questions

(1) What is the role of $NH_3 \cdot H_2O$-NH_4Cl buffer solution?

(2) Why do we need to add methyl red indicator in the experiment step and then add ammonia solution until the solution changes to a slight yellow?

实验 4　0.05mol·L⁻¹ EDTA 标准溶液的配制与标定

1. 实验目的

（1）掌握间接法配制 EDTA 标准溶液的方法。
（2）掌握 EDTA 标准溶液的标定方法。
（3）了解铬黑 T 指示剂终点的判断方法。

EDTA 标准溶液的标定

2. 实验原理

由于 EDTA 在水中的溶解度小，通常采用 EDTA 二钠盐配制标准溶液，也称为 EDTA 溶液。EDTA 二钠盐是一种白色结晶性粉末，分子量为 372.26。室温下，100mL 水中可溶解 11.1g EDTA 二钠盐。EDTA 标准溶液常采用间接法配制，即取适量 EDTA 二钠盐溶于水中摇匀，储存于硬质玻璃瓶中。以 ZnO 或 Zn 为基准物质，铬黑 T 或二甲酚橙作指示剂对其准确浓度进行标定。

3. 仪器和材料

（1）仪器　滴定管（25mL），锥形瓶（250mL），容量瓶（500mL），烧杯（50mL）。
（2）材料　EDTA 二钠盐，ZnO 基准物质，稀盐酸溶液，甲基红指示剂，氨试液，$NH_3·H_2O$-NH_4Cl 缓冲液，铬黑 T 指示剂。

4. 实验步骤

（1）0.05mol·L⁻¹ EDTA 标准溶液的配制　称取 9.5g EDTA 二钠盐，用适量水溶解后，转移到 500mL 容量瓶，继续用水稀释至刻度后摇匀。将配制好的溶液转移至硬质玻璃瓶中储存待用。

（2）0.05mol·L⁻¹ EDTA 标准溶液的标定　准确称取 800℃干燥至恒重的 ZnO 约 0.12g（精确至 ±0.0001g），加稀盐酸 3mL 使其溶解，再加入 25mL 蒸馏水及 1 滴甲基红指示剂，滴加氨试液至溶液呈现微黄色，再加入 25mL 蒸馏水、10mL $NH_3·H_2O$-NH_4Cl 缓冲液及 2 滴铬黑 T 指示剂，用 EDTA 溶液滴定至溶液由紫红色变成纯蓝色即为终点。计算公式：

$$c_{EDTA} = \frac{w_{ZnO} \times 1000}{M_{ZnO} V_{EDTA}}$$

5. 注意事项

（1）EDTA 二钠盐在水中溶解缓慢，需要不断搅拌或加热加速溶解。
（2）配制好的 EDTA 溶液应储存于硬质玻璃瓶中，否则 EDTA 会与瓶体中的金属离子反应，进而影响 EDTA 的浓度。
（3）滴定过程不能过快，尤其需注意接近滴定终点时应逐滴加入滴定剂，并不断振摇。

6. 思考

（1）加入 $NH_3·H_2O$-NH_4Cl 缓冲溶液的作用是什么？
（2）实验步骤中为什么需要加入甲基红指示剂后滴加氨试液至溶液呈现微黄色？

Exercises

4-1 Which form of EDTA is used in preparing a titrating solution? Why is a solution containing a metal ion buffered before titration with EDTA?

4-2 What are the characteristics of the complexes formed by EDTA and metal ions?

4-3 What conditions should the metallochromic indicators have?

4-4 Describe the blocking and ossification of indicators, and how to avoid them?

4-5 What are the methods of masking?

4-6 What is the coordination relationship between EDTA and most metal ions?
   A. 1 : 1　　　B. 1 : 2　　　C. 2 : 2　　　D. 2 : 1

4-7 Which is the EDTA type body directly coordinated with metal ions?
   A. H_6Y^{2+}　　　B. H_4Y　　　C. H_2Y^{2-}　　　D. Y^{4-}

4-8 The effective concentration of EDTA is related to acidity. How does it change with the increase of pH value of the solution?
   A. increase
   B. reduce
   C. be unchanged
   D. increase first and then decrease

4-9 What are the methods of coordination titration ?
   A. direct titration
   B. back titration
   C. replacement titration
   D. indirect titration

4-10 When the metal ion is titrated by EDTA, which one decreases with the increase of $[H^+]$?
   A. $\lg\alpha_{Y(H)}$
   B. $\lg K'_{MY}$
   C. $\alpha_{Y(H)}$
   D. $[HY]+[H_2Y]+\cdots+[H_6Y]$

4-11 Which is the condition for determining single metal ions in coordination titration analysis?
   A. $\lg(cK'_{MY}) \geq 8$　　B. $cK'_{MY} \geq 10^{-8}$　　C. $\lg(cK'_{MY}) \geq 6$　　D. $cK'_{MY} \geq 10^{-6}$

4-12 Which is the factor affecting the magnitude of the jump range of titration curve when the metal ions is titrated with EDTA ?
   A. concentration of EDTA
   B. concentration of metal ion
   C. concentration of the complex
   D. acid effect of EDTA

4-13 The complex formed by EDTA and metal ions is MY, and the complex formed by metal ions and indicators is MIn. What will happen when $K'_{MIn} > K'_{MY}$?
   A. Ossification
   B. Oxidative deterioration
   C. Blocking
   D. Masking

4-14 In the determination of Ca^{2+} in water by coordination titration, which is the common elimination method of Mg^{2+} interference?
   A. control the acidity of the solution
   B. coordination masking
   C. redox reaction masking
   D. precipitation masking

4-15 In the determination of Pb^{2+} by coordination titration, which is the simplest method to eliminate the interference of Ca^{2+} and Mg^{2+}?
   A. coordination masking
   B. redox reaction masking
   C. control the acidity of the solution
   D. precipitation masking

4-16 In the titration of Bi^{3+} with EDTA, which masking agent could be used to mask Fe^{3+}?
A. triethanolamine   B. KCN   C. oxalic acid   D. ascorbic acid

4-17 Calculate (a) $\alpha_{Y(H)}$ for EDTA at pH=5; (b) the percentage of $[Y^{4-}]$ in the total concentration of EDTA in this case.

$(10^{6.45}; 3.55 \times 10^{-5}\%)$

4-18 Calculate the conditional stability constant of the Mg^{2+}-EDTA complex at pH=6.0. Can the Mg^{2+} be titrated with the standard EDTA solution at pH=6.0 $[c(EDTA)=c(Mg^{2+})=0.01\ mol \cdot L^{-1}]$?

$(\lg K'_{MgY}=4.04;\ no)$

4-19 Calculate the allowed minimum pH value, in the mixed solution with Bi^{3+} and Ni^{2+} of $0.01\ mol \cdot L^{-1}$ titrated with EDTA. Can we use the method of controlling the acidity of the solution to make the separate titration of the two?

(0.6, 3.0; yes)

4-20 Calculate the minimum pH values for respectively titrated $0.01\ mol \cdot L^{-1}\ Fe^{2+}$ and $0.1\ mol \cdot L^{-1}\ Fe^{3+}$ with EDTA.

(5.2; 1.0)

Chapter 5　Redox Titration
第 5 章　氧化还原滴定法

 Study Guide　学习指南

Redox titration is a titration analysis method based on redox reactions. Compared with the acid-base titration method, redox titration method is much more complicated, because the mechanism of redox reaction is more complicated. Some reactions have a high degree of completeness, but reaction rate is very slow, sometimes measurement relationship between reactants is not determined due to occurrence of side reactions. Therefore, controlling appropriate conditions is particularly important in redox titration. Redox titration has a wide range of applications, many inorganic and organic substances can be directly or indirectly determined. This method uses oxidants or reducing agents as standard solutions, so it is divided into various titration methods such as potassium permanganate method, potassium dichromate method, and iodometric method. Each titration method has its characteristics and application range. This chapter focuses on the basic principles and applications of several commonly-used redox titration methods.

氧化还原滴定法是以氧化还原反应为基础的滴定分析法。与酸碱滴定法相比，氧化还原滴定法要复杂得多，因为氧化还原反应机理比较复杂。有些反应的完全程度很高，但反应速率很慢，有时由于副反应的发生使反应物之间没有确定的计量关系。因此，控制适当的条件在氧化还原滴定中显得尤为重要。氧化还原滴定的应用非常广泛，能直接或间接测定很多无机物和有机物。氧化还原滴定以氧化剂或还原剂作为标准溶液，据此分为高锰酸钾法、重铬酸钾法、碘量法等多种滴定方法。各种滴定方法都有其特点和应用范围。本章主要学习几种常用的氧化还原滴定法的基本原理和应用。

Section 1　Redox Reaction
第 1 节　氧化还原反应

1. Standard Electrode Potential and Conditional Electrode Potential

1. 标准电极电位和条件电极电位

In the redox reaction, strength of the oxidant and the

在氧化还原反应中，氧化剂和还原剂的强弱，可以用有关电对的电极电位

reducing agent can be measured by **electrode potential** (referred to as the potential) of the couples. The higher potential of the couples, the stronger oxidizing ability in oxidation state; the lower the potential, the stronger the reducing ability in the reduced state. An oxidant can oxidize a reducing agent with a lower potential than it; reducing agent can reduce an oxidizing agent with a higher potential than it. The electrode potential of the redox couple can be calculated by the Nernst formula. For example, the half-reaction of the following Ox/Red couples (omitting the charge of the ions):

（简称电位）来衡量。电对的电位越高，其氧化态的氧化能力越强；电位越低，其还原态的还原能力越强。氧化剂可以氧化电位比它低的还原剂；还原剂可以还原电位比它高的氧化剂。氧化还原电对的电极电位可用能斯特公式求得。例如，Ox/Red 的半反应：

$$Ox + ne \rightleftharpoons Red$$

The Nernst formula of the electric counter electrode potential is

电对电极电位的能斯特公式为

$$E_{Ox/Red} = E^{\ominus}_{Ox/Red} + \frac{RT}{nF} \ln \frac{a_{Ox}}{a_{Red}} \tag{5.1}$$

$E_{Ox/Red}$— Oxidation number (Ox) / Reduction number (Red) couples electrode potential;
$E^{\ominus}_{Ox/Red}$— Standard Electrode Potential (SEP);
a_{Ox}, a_{Red}—Activity in the oxidation and reduction, the ion activity is equal to the concentration c times the activity coefficient γ, $a=\gamma c$;
R—Molar gas constant, 8.314 J·mol^{-1}·K^{-1};
T—Thermodynamic temperature, K;
F—Faraday constant, 96485 C·mol^{-1};
n—number of electrons transferred in half-reaction.
Substituting the above data into formula (5.1), it can be obtained at 25 ℃

式中：
$E_{Ox/Red}$——氧化态（Ox）、还原态（Red）电对的电极电位；
$E^{\ominus}_{Ox/Red}$——标准电极电位；
a_{Ox}, a_{Red}——氧化态及还原态的活度，离子的活度等于浓度 c 乘以活度系数 γ，$a=\gamma c$；
R——摩尔气体常数，8.314J·mol^{-1}·K^{-1}；
T——热力学温度，K；
F——法拉第常数，96485 C·mol^{-1}；
n——半反应中电子的转移数。
将以上数据代入式（5.1），25℃时得：

$$E_{Ox/Red} = E^{\ominus}_{Ox/Red} + \frac{0.059V}{n} \lg \frac{a_{Ox}}{a_{Red}} \tag{5.2}$$

It can be seen from formula (5.2) that the electrode potential of the couples is related to the activity in the oxidized and reduced states in the solution. When $a_{Ox}=a_{Red}=1$, $E_{Ox/Red}= E^{\ominus}_{Ox/Red}$, electrode potential is equal to standard electrode potential at this time. Standard electrode potential means that at a certain temperature (usually 25℃), each component in the redox half reaction is in a standard state, that is, the degree of

从式（5.2）中可见，电对的电极电位与存在于溶液中氧化态和还原态的活度有关。当 $a_{Ox}=a_{Red}=1$ 时，$E_{Ox/Red}=E^{\ominus}_{Ox/Red}$，这时的电极电位等于标准电极电位。所谓标准电极电位是指在一定温度下（通常为25℃），氧化还原半反应中各组分都处于标准状态，即离子或分子的活度等于1mol·L^{-1}，反应中若有

activity of ions or molecules is equal to 1mol·L^{-1}. If gas participates in the reaction, then its partial voltage is equal to the electrode potential at 100kPa. $E_{Ox/Red}^{\ominus}$ only changes with temperature.

For simplicity, the effect of ionic strength in the solution is ignored, and concentration of solution is usually used instead of activity. However, in actual work, effect of ionic strength in the solution cannot be ignored. More importantly, when composition of solution changes, existence form of oxidation-reduction couples also changes accordingly, which causes electrode potential to change. In this case, when using the Nernst formula to calculate electrode potential of couples, calculation result is very different from the actual situation if the standard potential is still used and effect of ionic strength is not considered. For example, the Nernst formula is used to calculate the potential of the Fe^{3+} and Fe^{2+} system in HCl solution as follows:

$$E_{Fe^{3+}/Fe^{2+}} = E_{Fe^{3+}/Fe^{2+}}^{\ominus} + 0.059V \lg \frac{a_{Fe^{3+}}}{a_{Fe^{2+}}} \tag{5.3}$$

$$E_{Fe^{3+}/Fe^{2+}} = E_{Fe^{3+}/Fe^{2+}}^{\ominus} + 0.059V \lg \frac{\gamma_{Fe^{3+}}[Fe^{3+}]}{\gamma_{Fe^{2+}}[Fe^{2+}]} \tag{5.4}$$

On the other hand, in addition to Fe^{3+} and Fe^{2+} in HCl solution, there are also other existing forms of Trivalent iron $Fe(OH)^{2+}$, $FeCl^{2+}$, $FeCl_2^+$, $FeCl_4^-$, $FeCl_6^{3-}$ and other existing forms of divalent iron $Fe(OH)^+$, $FeCl^+$, $FeCl_3^-$, $FeCl_4^{2-}$ in solution. If $c_{Fe(III)}$、$c_{Fe(II)}$ are represented the total concentration of various existing forms of trivalent iron and divalent iron in solution, then

$$\alpha_{Fe^{3+}} = \frac{c_{Fe(III)}}{[Fe^{3+}]}$$

$$\alpha_{Fe^{2+}} = \frac{c_{Fe(II)}}{[Fe^{2+}]}$$

$\alpha_{Fe^{3+}}$ and $\alpha_{Fe^{2+}}$ are the side reaction coefficients of Fe^{3+} and Fe^{2+} in the HCl solution respectively. Substituted into equation (5.4), we know

为了简化起见，通常忽略溶液中离子强度的影响，就以溶液的浓度代替活度进行计算。但在实际工作中，溶液中离子强度的影响不能忽视，更重要的是当溶液组成改变时，电对的氧化态和还原态的存在形式也随之改变，因而引起电极电位的变化，在这种情况下，用能斯特公式计算有关电对的电极电位时，若仍采用标准电位，不考虑离子强度的影响，其计算结果与实际情况相差很大。例如，用Nernst方程计算Fe^{3+}和Fe^{2+}体系在HCl溶液中的电势如下：

另一方面，在HCl溶液中除Fe^{3+}、Fe^{2+}外，三价铁还有$Fe(OH)^{2+}$、$FeCl^{2+}$、$FeCl_2^+$、$FeCl_4^-$、$FeCl_6^{3-}$等存在形式，而二价铁也还有$Fe(OH)^+$、$FeCl^+$、$FeCl_3^-$、$FeCl_4^{2-}$等存在形式。若用$c_{Fe(III)}$、$c_{Fe(II)}$分别表示溶液中三价铁Fe(III)和二价铁Fe(II)各种存在形式的总浓度，$\alpha_{Fe^{3+}}$及$\alpha_{Fe^{2+}}$分别是HCl溶液中Fe^{3+}、Fe^{2+}的副反应系数，则得式（5.5）。因为Fe^{3+}、Fe^{2+}的总浓度$c_{Fe(III)}$、$c_{Fe(II)}$是已知

$$E_{Fe^{3+}/Fe^{2+}} = E^{\ominus}_{Fe^{3+}/Fe^{2+}} + 0.059\text{V} \lg \frac{\gamma_{Fe^{3+}} \cdot \alpha_{Fe^{2+}} \cdot c_{Fe(III)}}{\gamma_{Fe^{2+}} \cdot \alpha_{Fe^{3+}} \cdot c_{Fe(II)}} \tag{5.5}$$

Because the total concentrations [$c_{Fe(III)}$ and $c_{Fe(II)}$] of Fe^{3+} and Fe^{2+} are known, α 和 γ are fixed values under certain conditions and can be incorporated into the constant, so we may rewrite equation (5.5) as

的，α 和 γ 在一定条件下为固定值，可以并入常数项中，为此式（5.5）可改写为式（5.6）。

$$E_{Fe^{3+}/Fe^{2+}} = E^{\ominus}_{Fe^{3+}/Fe^{2+}} + 0.059\text{V} \lg \frac{\gamma_{Fe^{3+}} \cdot \alpha_{Fe^{2+}}}{\gamma_{Fe^{2+}} \cdot \alpha_{Fe^{3+}}} + 0.059\text{V} \lg \frac{c_{Fe(III)}}{c_{Fe(II)}} \tag{5.6}$$

$$E^{\ominus'}_{Fe^{3+}/Fe^{2+}} = E^{\ominus}_{Fe^{3+}/Fe^{2+}} + 0.059\text{V} \lg \frac{\gamma_{Fe^{3+}} \cdot \alpha_{Fe^{2+}}}{\gamma_{Fe^{2+}} \cdot \alpha_{Fe^{3+}}}$$

Rewrite equation (5.6) as

$$E_{Fe^{3+}/Fe^{2+}} = E^{\ominus'}_{Fe^{3+}/Fe^{2+}} + 0.059\text{V} \lg \frac{c_{Fe(III)}}{c_{Fe(II)}} \tag{5.7}$$

In the formula, $E^{\ominus'}_{Fe^{3+}/Fe^{2+}}$ is called a **conditional electrode potential**. It indicates that under a certain medium condition, total concentration of the oxidization and reduction number is $1\text{mol} \cdot \text{L}^{-1}$, or concentration ratio of them is 1, sactual potential is corrected after correction of various external factor. The conditional electrode potential reflects ionic strength and overall effect of various side reactions, and it is constant under certain conditions.

式中，$E^{\ominus'}_{Fe^{3+}/Fe^{2+}}$ 称为条件电极电位。它表示在一定介质条件下氧化态和还原态的总浓度都为 $1\text{mol} \cdot \text{L}^{-1}$ 或二者浓度比值为 1 时校正了各种外界因素影响后的实际电位，条件电极电位反映了离子强度与各种副反应影响的总结果，在一定条件下为常数。

Discussion

You can compare values of standard electrode potential with conditional electrode potential of the same couples by yourself. The difference between two is quite large. Therefore, conditional electrode potential should be used as much as possible during calculating potential of redox reaction. When electrode potential data under the same conditions are not available, conditional electrode potentials with similar conditions can be used, so that the obtained results are closer to the actual situation than before.

2. Completeness of Redox Reaction

In redox titration analysis, it is required that redox reaction proceeds as completely as possible, and completeness of the reaction is measured by its equilibrium constant.

2. 氧化还原反应进行的程度

在氧化还原滴定分析中，要求氧化还原反应进行得越完全越好，而反应的完全程度是以它的平衡常数大小来衡

The equilibrium constant can be obtained according to the Nernst formula and conditional electrode potential or standard electrode potential related to the couples. If conditional potential is quoted, the conditional equilibrium constant K' is obtained, which can better explain the actual progress of reaction. Redox reaction equation is:

量。平衡常数可以根据能斯特公式和有关电对的条件电极电位或标准电极电位求得。若引用条件电位，求得的是条件平衡常数 K'，它更能说明反应实际进行的程度。

$$n_2 O_{X1} + n_2 Red_2 \rightleftharpoons n_2 Red_1 + n_1 O_{X2}$$

Electrode potential of two couples is

$$Ox_1 + n_1 e \rightleftharpoons Red_1 \qquad E_1 = E_1^{\ominus'} + \frac{0.059\text{V}}{n_1} \lg \frac{c_{Ox_1}}{c_{Red_1}}$$

$$Ox_2 + n_2 e \rightleftharpoons Red_2 \qquad E_2 = E_2^{\ominus'} + \frac{0.059\text{V}}{n_2} \lg \frac{c_{Ox_2}}{c_{Red_2}}$$

When reaction reaches equilibrium, $E_1 = E_2$, then,

$$E_1^{\ominus'} + \frac{0.059\text{V}}{n_1} \lg \frac{c_{Ox_1}}{c_{Red_1}} = E_2^{\ominus'} + \frac{0.059\text{V}}{n_2} \lg \frac{c_{Ox_2}}{c_{Red_2}} \tag{5.8}$$

When reaction reaches equilibrium,

$$K' = \frac{(c_{Red_1})^{n_2} \cdot (c_{Ox_2})^{n_1}}{(c_{Ox_1})^{n_2} \cdot (c_{Red_2})^{n_1}} \tag{5.9}$$

$$\lg K' = \frac{(E_1^{\ominus'} - E_2^{\ominus'}) n_1 n_2}{0.059\text{V}} \tag{5.10}$$

It is known that the greater difference between the conditional potentials of the two electric pairs, the greater conditional equilibrium constant K in oxidation-reduction reaction, and reaction is more completely. For the titration reaction, completeness of reaction should be above 99.9%. According to the formula (5.10), the quantitative conditions for the oxidation-reduction titration reaction can be known. If $n_1 = n_2 = 1$, when titrating reaches stoichiometric point, it is required:

可见，两电对的条件电位相差越大，氧化还原反应的条件平衡常数 K 就越大，反应进行得越完全。对于滴定反应来说，反应的完全程度应当在 99.9% 以上，根据式（5.9）可以得到氧化还原滴定反应定量进行的条件。

$$\frac{c_{Ox_2}}{c_{Red_2}} \geqslant 10^3, \quad \frac{c_{Red_1}}{c_{Ox_1}} \geqslant 10^3$$

$$K' = \frac{c_{Red_1} \cdot c_{Ox_2}}{c_{Ox_1} \cdot c_{Red_2}} \geqslant 10^6$$

$$E_1^{\ominus'} - E_2^{\ominus'} = \frac{0.059\text{V}}{n_1 n_2} \lg K' \geqslant 0.059\text{V} \times 6 \approx 0.4\text{V}$$

Generally, redox reaction should be carried out quantitatively, when reaction reaches equilibrium, $\lg K' \geqslant 6$, $E_1^\ominus - E_2^\ominus \geqslant 0.4\text{V}$, redox reaction can be applied to titration analysis. In redox titration, there are many strong oxidants and strong reducing agents as titrants, and change the electrode potential of the couples by controlling conditions of the medium to achieve this requirement.

3. Factors Affecting the Rate of Redox Reaction

According to the conditional potential of the electron pair, direction and completeness of redox reaction can be judged, but this case can only show possibility of reaction and not show speed of reaction. In titration analysis, redox reaction must be carried out quantitatively and quickly. Therefore, not only possibility of the reaction from an equilibrium point but also rate of reaction should be considered. The factors that affect the redox reaction rate are the following.

(1) Reactant concentration   According to the law of mass action, reaction rate is proportional to product of reactant concentration. However, many redox reactions are carried out step by step, and the rate of entire reaction is determined by the slowest step. Therefore, influence of concentration on the reaction rate cannot be judged by coefficient of each reactant in the total redox reaction equation. However, in general, increasing concentration of the reaction substance can speed up reaction rate. For example, potassium dichromate reacts and potassium iodide in an acidic solution.

$$Cr_2O_7^{2-} + 6I^- + 14H^+ \rightleftharpoons 2Cr^{3+} + 3I_2 + 7H_2O$$

When reaction rate is slow, increasing concentration of I^- and H^+ can accelerate reaction. The experiment proves that under the condition of $[H^+]=0.4\ \text{mol}\cdot\text{L}^{-1}$, overdose of KI is about 5 times , and reaction can be completed when it is left for 5 minutes, but acidity cannot be too large, otherwise it will promote oxidation of I^- by oxygen in the air, which causes analytical error.

一般的，氧化还原反应要定量地进行，则该反应达到平衡时，其$\lg K' \geqslant 6$，$E_1^\ominus - E_2^\ominus \geqslant 0.4\text{V}$，这样的氧化还原反应才能应用于滴定分析。在氧化还原滴定中，有很多强的氧化剂和较强的还原剂可用作滴定剂，还可以控制介质条件来改变电对的电极电位，以达到此要求。

3. 影响氧化还原反应速率的因素

根据有关电对的条件电位，可以判断氧化还原反应的方向和完全程度，但这只能说明反应发生的可能性，不能表明反应速率的快慢。而在滴定分析中，要求氧化还原反应必须定量、迅速地进行，所以对于氧化还原反应除了要从平衡观点来了解反应的可能性外，还应考虑反应的速率。影响氧化还原反应速率的因素主要有以下几个方面。

（1）反应物浓度   根据质量作用定律，反应速率与反应物浓度的乘积成正比。但许多氧化还原反应是分步进行的，整个反应的速率由最慢的一步决定，因此不能笼统地按总的氧化还原反应方程式中各反应物的系数判断其浓度对反应速率的影响。不过一般情况下，增加反应物质的浓度均可以加快反应速率。例如，在酸性溶液中重铬酸钾和碘化钾反应。

反应速率较慢，提高I^-和H^+的浓度，可加速反应。实验证明，在$[H^+]=0.4\ \text{mol}\cdot\text{L}^{-1}$条件下，KI过量约5倍，放置5min，反应即可进行完全，但酸度不能太大，否则将促使空气中的氧对I^-的氧化，造成分析误差。

(2) **Temperature** The effect of temperature on reaction rate is also very complicated. For most reactions, high temperature can speed up reaction rate. Generally, reaction rate increases by 2 to 3 times for every 10℃ increase in temperature. For example, reaction of potassium permanganate and oxalic acid:

$$2MnO_4^- + 5C_2O_4^{2-} + 16H^+ \rightleftharpoons 2Mn^{2+} + 10CO_2 + 8H_2O$$

Reaction rate is very slow at normal temperature. If the temperature is controlled at 75-85 ℃, reaction rate will increase significantly. However, high temperature is not beneficial for all redox reactions. As for reaction of $K_2Cr_2O_7$ and KI described above, if heating is used to accelerate reaction rate, I_2 generated will volatilize and cause losses. For another example, if the heating temperature of oxalic acid solution is too high or time is too long, oxalic acid will decompose and cause analytical error. Some reducing substances, such as Fe^{2+} and Sn^{2+} will be more easily oxidized by oxygen in the air due to heating, which also cause errors in analysis results.

(3) **Catalyst** One of the effective methods to speed up reaction rate is using catalyst. For example, reaction of $KMnO_4$ and $H_2C_2O_4$ in the acidic solution, even if temperature of solution is increased, $KMnO_4$ still fades slowly in the initial stage of titration. If a small amount of Mn^{2+} is added, reaction can go quickly, and Mn^{2+} acts as a catalyst. In practice, catalyst Mn^{2+} may not be added, because in acidic medium, one of products of reaction between MnO_4^- and $C_2O_4^-$ is Mn^{2+}. With reaction is going, Mn^{2+} concentration gradually increases, and reaction rate will become faster and faster. This kind of reaction which is catalyzed by the product itself is called **autocatalytic reaction**.

(4) **Induced reaction** Some redox reactions do not occur or proceed very slowly under normal circumstances, but will be promoted when another reaction proceeds. This kind of redox reaction is promoted by the occurrence of another redox reaction, which is called **induced reaction**. For example, in an acidic solution, reaction rate of

（2）温度　温度对反应速率的影响也是很复杂的，温度的升高对于大多数反应来说，可以加快反应速率。通常温度每升高10℃，反应速率增加2～3倍。例如，高锰酸钾与草酸的反应，在常温下反应速率很慢，若温度控制在75～85℃时，反应速率显著提高。

但是，提高温度并不是对所有氧化还原反应都是有利的。上面介绍的$K_2Cr_2O_7$和KI的反应，若用加热方法来加快反应速率，则生成的I_2反而会挥发而引起损失。又如，草酸溶液加热温度过高或时间过长，草酸将分解而引起误差，有些还原性物质如Fe^{2+}、Sn^{2+}等会因加热而更容易被空气中的氧所氧化，也造成分析结果的误差。

（3）催化剂　使用催化剂是加快反应速率的有效方法之一。例如，在酸性溶液中$KMnO_4$与$H_2C_2O_4$的反应，即使将溶液的温度升高，在滴定的最初阶段，$KMnO_4$褪色仍很慢，若加入少许Mn^{2+}，反应就能很快进行，这里Mn^{2+}起催化剂作用。实际应用中也可不外加催化剂Mn^{2+}，因为在酸性介质中，MnO_4^-与$C_2O_4^-$反应的生成物之一就是Mn^{2+}，随着反应的进行，Mn^{2+}浓度逐渐增大，反应速率也将越来越快。这种由于生成物本身引起催化作用的反应称为自动催化反应。

（4）诱导反应　有些氧化还原反应在通常情况下并不发生或进行极慢，但在另一反应进行时会促进这一反应的发生。这种由于一个氧化还原反应的发生促进另一氧化还原反应进行，称为诱导反应。例如，在酸性溶液中，$KMnO_4$

KMnO₄ to oxidize Cl⁻ is extremely slow. When Fe^{2+} is in the solution, reaction of KMnO₄ to oxidize Fe^{2+} will accelerate reaction of KMnO₄ to oxidize Cl⁻. Here Fe^{2+} is called inducer, MnO_4^- is called bearer, and Cl⁻ is called attractant.

Induced reaction is different from catalytic reaction. In the catalytic reaction, catalyst returns to its original state after participating in reaction. In induced reaction, the inducer becomes other substances after reaction and attractant also participates in reaction, which increases consumption of the effector. Therefore, titrate Fe^{2+} with KMnO₄, when Cl⁻ is present, which will increase consumption of KMnO₄ solution and cause deviation in measurement results. If KMnO₄ method is used to measure Fe^{2+} in HCl media, mixed solution of $MnSO_4$-H_3PO_4-H_2SO_4 should be added to solution to prevent reduction of Cl⁻ on MnO_4^- to obtain correct titration results.

Section 2　Redox Titration
第 2 节　氧化还原滴定

The titration curve (fig.5-1) that is shown here is for the analysis of a 10.00mL portion of a 0.0100mol·L⁻¹ Fe^{2+} solution using 0.0050 mol·L⁻¹ Ce^{4+} as the titrant.

图 5-1 的曲线是以 0.0050mol·L⁻¹ 的 Ce^{4+} 滴定 10.00mL 0.0100mol·L⁻¹ Fe^{2+} 的滴定曲线。

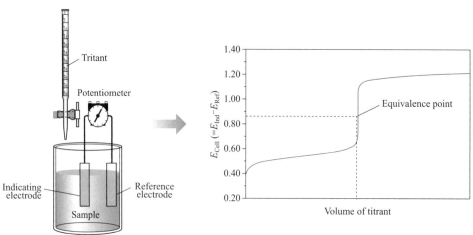

Fig.5-1　A typical system used to perform a redox titration that is monitored through the use of potential measurements

图 5-1　典型的氧化还原滴定反应装置及滴定曲线（响应信号用电位测量法测定）

1. Redox Titration Curve

Similar to other titration analysis methods, during the redox titration process, with addition of titrant, concentration of oxidant and reducing agent in the solution gradually changes, and potential of electrical pair also changes continuously. This change can be described by titration curve. Redox titration curve can be drawn by measuring experimental data. If two couples can reversible in reaction, titration curve can be drawn from conditional potential values of two couples according to the Nernst formula.

For example, now we discuss some changes between amount of standard solution and amount of electrode potential during titrating 20.00mL 0.1000mol·L^{-1} FeSO$_4$ with standard solution of 0.1000mol·L^{-1} Ce(SO$_4$)$_2$ in 1mol·L^{-1} H$_2$SO$_4$ solution,

Formula:

$$Ce^{4+} + Fe^{2+} \xrightleftharpoons[]{1mol·L^{-1} H_2SO_4} Ce^{3+} + Fe^{3+}$$

Conditional electrode potential of two couples

$$Fe^{3+} + e \longrightarrow Fe^{2+} \qquad E^{\ominus'}_{Fe^{3+}/Fe^{2+}} = 0.68V$$

$$Ce^{4+} + e \longrightarrow Ce^{3+} \qquad E^{\ominus'}_{Ce^{4+}/Ce^{3+}} = 1.44V$$

Before titration, $c_{Fe^{2+}}=0.1000$mol·L^{-1}. After the titration starts, with instillation of Ce^{4+} standard solution, Fe^{2+} concentration gradually decreases, and Fe^{3+} concentration gradually increases, meanwhile Ce^{4+} is reduced to Ce^{3+}. Therefore, there are two pairs in the system at the same time from the beginning to the end of titration. When equilibrium is reached at any point in the titration process, the potentials of two electric pairs are equal, that is,

$$E = E_{Fe^{3+}/Fe^{2+}} = E_{Ce^{4+}/Ce^{3+}}$$

$$E_{Fe^{3+}/Fe^{2+}} = E^{\ominus'}_{Fe^{3+}/Fe^{2+}} + 0.059V \lg \frac{[Fe^{3+}]}{[Fe^{2+}]}$$

$$E_{Ce^{4+}/Ce^{3+}} = E^{\ominus'}_{Ce^{4+}/Ce^{3+}} + 0.059V \lg \frac{[Ce^{4+}]}{[Ce^{3+}]}$$

1. 氧化还原滴定曲线

与其他滴定分析法相似,在氧化还原滴定过程中,随着滴定剂的加入,溶液中氧化剂和还原剂的浓度逐渐改变,有关电对的电位也随之不断变化,这种变化可用滴定曲线来描述。氧化还原滴定曲线可以由实验数据而绘出,若反应中两电对都是可逆的,就可以根据能斯特方程,由两电对的条件电位值计算得到。

现以在1mol·L^{-1} H$_2$SO$_4$溶液中,用0.1000mol·L^{-1} Ce(SO$_4$)$_2$标准溶液滴定20.00mL 0.1000mol·L^{-1} FeSO$_4$为例,讨论滴定过程中标准溶液用量和电极电位之间量的变化情况。两个电对的条件电位:$E^{\ominus'}_{Fe^{3+}/Fe^{2+}}$=0.68V,$E^{\ominus'}_{Ce^{4+}/Ce^{3+}}$=1.44V。

滴定开始前,$c_{Fe^{2+}}$=0.1000mol·L^{-1}。滴定开始后,随着Ce^{4+}标准溶液的滴入,Fe^{2+}的浓度逐渐减小,Fe^{3+}的浓度逐渐增加;滴入的Ce^{4+}被还原为Ce^{3+}。所以,从滴定开始至滴定结束,体系中就同时存在两个电对,在滴定过程中任何一点达到平衡时,两电对的电位相等。

Therefore, at different stages of the titration, an easy-to-calculate pair of electricity can be selected, and system's electricity during titration process can be calculated according to the Nernst formula. Calculation method of potential value for each titration point is as follows.

(1) Titration starting until stoichiometric point   Because almost all Ce^{4+} added is reduced to Ce^{3+}, when equilibrium is reached, concentration of Ce^{4+} is extremely small, which is not easy to obtain directly. But if titration fraction is known, $c_{Ce^{4+}}/c_{Ce^{3+}}$ value is determined. At this time, E can be calculated by Fe^{3+}/Fe^{2+} pair. For example, when 99.9% Fe^{2+} is titrated,

$$\frac{c_{Fe^{3+}}}{c_{Fe^{2+}}} = \frac{99.9}{0.1} \approx 1000$$

$$E = E_{Fe^{3+}/Fe^{2+}} = E^{\ominus'}_{Fe^{3+}/Fe^{2+}} + 0.059V \lg\frac{c_{Fe^{3+}}}{c_{Fe^{2+}}} = (0.68 + 0.059 \times 3)V = 0.86V$$

Before stoichiometric point, potential value at each titration point can be calculated in the same way.

(2) At stoichiometric point

$$E = \frac{n_1 E^{\ominus'}_{Ce^{4+}/Ce^{3+}} + n_2 E^{\ominus'}_{Fe^{3+}/Fe^{2+}}}{n_1 + n_2} = \frac{1.44V + 0.68V}{2} = 1.06V$$

(3) After stoichiometric point   All Fe^{2+} is completely oxidized to Fe^{3+}, and concentration of Fe^{2+} is extremely small, which is not easy to obtain directly. If excess Ce^{4+} was added at this time, it is more convenient to calculate E value by $c_{Ce^{4+}}/c_{Ce^{3+}}$ couples.

$$E_{Ce^{4+}/Ce^{3+}} = E^{\ominus'}_{Ce^{4+}/Ce^{3+}} + 0.059V \lg\frac{[Ce^{4+}]}{[Ce^{3+}]}$$

For example, when Ce^{4+} is excess 0.1%, solution potential is

$$E_{Ce^{4+}/Ce^{3+}} = E^{\ominus'}_{Ce^{4+}/Ce^{3+}} + 0.059V \lg\frac{0.1}{100} = 1.26V$$

Potential value of each titration point after stoichiometric point can be calculated in the same way (equations are in fig.5-2).

During titration process, potential calculation results of different titration points are listed in table 5-1. Titration curve drawn is shown in figure 5-3.

因此，在滴定的不同阶段，可选用便于计算的电对，按能斯特方程计算滴定过程中体系的电位值。

（1）滴定开始至化学计量点前　因加入的 Ce^{4+} 几乎全部被还原成 Ce^{3+}，达到平衡时，Ce^{4+} 的浓度极小，不易直接求得。但如果知道了滴定分数，$c_{Ce^{4+}}/c_{Ce^{3+}}$ 值就确定了，这时可以利用 Fe^{3+}/Fe^{2+} 电对来计算 E 值。

在化学计量点前各滴定点的电位值可按同法计算。

（2）化学计量点时

（3）化学计量点后　此时，Fe^{2+} 几乎全部被氧化为 Fe^{3+}，Fe^{2+} 的浓度极小，不易直接求得。此时加入了过量的 Ce^{4+}，利用 $c_{Ce^{4+}}/c_{Ce^{3+}}$ 电对计算 E 值较方便。

化学计量点过后各滴定点的电位值，可按同法计算，计算式见图5-2。

滴定过程中，不同滴定点的电位计算结果列于表5-1，由此绘制的滴定曲线如图5-3所示。

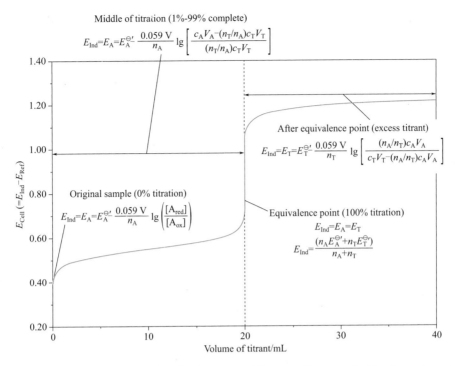

Fig.5-2 Four general regions in a redox titration curve (The specific curve that is shown here is for the analysis of a 10.00mL sample of 0.0100 mol·L^{-1} Fe^{2+} with 0.0050 mol·L^{-1} Ce^{4+} as the titrant.)

图 5-2 氧化还原滴定曲线的 4 个阶段（该曲线是 0.0050mol·L^{-1} 的 Ce^{4+} 滴定 10.00mL 0.0100mol·L^{-1} Fe^{2+} 的滴定曲线。）

Table 5-1 Titrate 20.00mL 0.1000mol·L^{-1}Fe^{3+} solution with 0.1000mol·L^{-1}Ce(SO$_4$)$_2$ in 1 mol·L^{-1} H$_2$SO$_4$ solution

表 5-1 在 1 mol·L^{-1}H$_2$SO$_4$ 溶液中，用 0.1000 mol·L^{-1}Ce(SO$_4$)$_2$ 溶液滴定 20.00 mL 0.1mol·L^{-1}Fe^{3+} 溶液

Ce(SO$_4$)$_2$		potential/V
V/mL	a/%	
1.00	5.0	0.60
2.00	10.0	0.62
4.00	20.0	0.64
8.00	40.0	0.67
10.00	50.0	0.68
12.00	60.0	0.69
18.00	90.0	0.74
19.80	99.0	0.80
19.98	99.9	0.86 ⎫
20.00	100.0	1.06 ⎬ titration jump
20.02	100.1	1.26 ⎭
22.00	110.0	1.38
30.00	150.0	1.42
40.00	200.0	1.44

As shown in table 5-1, when Ce^{4+} standard solution is dropped into 50%, the potential is equal to conditional electrode potential of the reducing agent pair; when Ce^{4+} standard solution is dropped into 200%, the potential is equal to the conditional electrode potential of the oxidant couple; when titration is from 99.9% to 100.1%, electrode potential change range is 1.26V−0.86V=0.4V(fig. 5-3), that is, potential jump of titration curve is 0.4V, which provides a basis for judging possibility of oxidation-reduction reaction titration and selecting indicator. In reaction of Ce^{4+} titration of Fe^{2+}, electron transfer number of two couples is one, the potential of stoichiometric point (1.06V) is exactly in the middle of titration jump (0.86-1.26V), and the entire titration curve is generally symmetrical. The length of the redox titration curve jump is related to the difference in conditional electrode potential of two pairs of oxidant and reductant. Conditional electrode potentials of two electric pairs are quite different, and titration jump is longer. On the contrary, titration jump is shorter.

For a symmetrical couple with $n_1 = n_2$ (meaning the same oxidation and reduction coefficients), the potential at stoichiometric point is in the middle of titration jump, and titration curve before and after stoichiometric point is generally symmetrical.

从表5-1可见，当Ce^{4+}标准溶液滴入50%时的电位等于还原剂电对的条件电极电位；当Ce^{4+}标准溶液滴入200%时的电位等于氧化剂电对的条件电极电位。如图5-3所示，滴定99.9%～100.1%时电极电位变化范围为1.26V−0.86V=0.4V，即滴定曲线的电位突跃是0.4V，这为判断氧化还原反应滴定的可能性和选择指示剂提供了依据。Ce^{4+}滴定Fe^{2+}的反应中，两电对电子转移数都是1，化学计量点的电位（1.06V）正好处于滴定突跃中间（0.86～1.26V），整个滴定曲线基本对称。氧化还原滴定曲线突跃的长短与氧化剂还原剂两电对的条件电极电位的差值大小有关。两电对的条件电极电位相差较大，滴定突跃较长；反之，其滴定突跃就较短。

对于$n_1=n_2$的对称电对（指氧化态与还原态系数相同），其化学计量点时的电位就处于滴定突跃的中间，化学计量点前后的滴定曲线基本上是对称的。

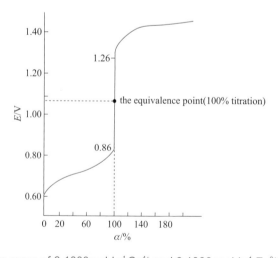

Fig. 5-3 Titration curve of 0.1000mol·L^{-1} Ce^{4+} and 0.1000 mol·L^{-1} Fe^{2+}(1mol·L^{-1}H_2SO_4)

图5-3 在1mol·L^{-1}的H_2SO_4介质中，用0.1000mol·L^{-1}的Ce^{4+}滴定0.1000mol·L^{-1}Fe^{2+}的滴定曲线

For redox reaction of $n_1 \neq n_2$, titration curve is asymmetric around the stoichiometric point, and potential of stoichiometric point is not at the center of titration jump, but it is biased towards the electric pair with more electron gains and losses.

It should be noted that for irreversible pairs (such as MnO_4^-/Mn^{2+}, $K_2Cr_2O_7/Cr^{3+}$, $S_4O_6^{2-}/S_2O_3^{2-}$ etc.), their potential calculations do not follow the Nernst formula, so calculated titration curve is significantly different from actual titration curve. Titration curves of irreversible redox systems are determined by some experiments.

2. Determination of the Redox Titration End Point

In redox titration, in addition to potentiometric method to determine end point, indicator is usually used to indicate the end point of titration. There are three types of commonly-used indicators in redox titration.

(1) **Self indicator** In the process of redox titration, some standard solutions or measured substance have their own color, so there is no need to add an indicator when titrating. The color change of itself plays the role of an indicator, which is called a self-indicator. For example, titrate $FeSO_4$ solution with $KMnO_4$ standard solution:

$$MnO_4^- + 5Fe^{2+} + 8H^+ \longrightarrow Mn^{2+} + 5Fe^{3+} + 4H_2O$$

Because $KMnO_4$ itself has a fuchsia color, and Mn^{2+} is almost colorless, when titration reaches stoichiometric point, slight excess $KMnO_4$ causes solution to appear pink, indicating that titration end point reaches. The experiment proves that the pink color of solution can be observed when concentration of $KMnO_4$ is about 2×10^{-6} mol·L^{-1}.

(2) **Starch indicator** Reaction between soluble starch and free iodine togenerate a dark blue complex is an exclusive reaction. When I_2 is reduced to I^-, blue disappears; when I^- is oxidized to I_2, blue appears. When concentration of I_2 is 2×10^{-6} mol·L^{-1}, blue can

be seen, and reaction is extremely sensitive. So starch is exclusive indicator of iodine method.

(3) Redox indicator   This type of indicator is an organic compound with its own redox properties. Redox can occur during redox titration. Its oxidization state and reduction state have different colors, which can indicate end point of redox titration. Oxidation and reduction states of indicator are represented by Ox and Red respectively. The redox half reaction is as follows:

$$Ox + ne \rightleftharpoons Red$$

According to the Nernst formula, we know that

$$E_{In} = E_{In}^{\ominus'} + \frac{0.059V}{n} \lg \frac{c_{Ox}}{c_{Red}}$$

In the formula, $E_{In}^{\ominus'}$ is conditional electrode potential of indicator. With change of potential of titration system, concentration ratio of oxidation state and reduction state of indicator also changes. So does discoloration of acid-base indicator. Potential range of discoloration of redox indicator is

$$E_{In}^{\ominus'} \pm \frac{0.059V}{n}$$

It is noted that different indicators have different $E_{In}^{\ominus'}$, and the same indicator has different $E_{In}^{\ominus'}$ in different media. Conditioned electrode potentials of some important redox indicators are listed in table 5-2. When selecting indicator, conditioned electrode potential of redox indicator should be as consistent as possible with potential of stoichiometric point to reduce error of titration end point. Redox indicator is a universal indicator for redox titration. The principle for selection is that conditional potential of indicator should be within titration jump.

色，反应极灵敏。因而淀粉是碘量法的专属指示剂。

（3）氧化还原指示剂   这类指示剂是本身具有氧化还原性质的有机化合物。在氧化还原滴定过程中能发生氧化还原反应，而它的氧化态和还原态具有不同的颜色，因而可指示氧化还原滴定终点。现以 Ox 和 Red 分别表示指示剂的氧化态和还原态：

根据能斯特方程得

式中，$E_{In}^{\ominus'}$ 为指示剂的条件电极电位。随着滴定体系电位的改变，指示剂氧化态和还原态的浓度比也发生变化，因而使溶液的颜色发生变化，同酸碱指示剂的变色情况相似，氧化还原指示剂变色的电位范围是：

必须注意，指示剂不同，其 $E_{In}^{\ominus'}$ 不同，同一种指示剂在不同的介质中，其 $E_{In}^{\ominus'}$ 也不同。表 5-2 列出了一些重要的氧化还原指示剂的条件电极电位。在选择指示剂时，应使氧化还原指示剂的条件电极电位尽量与反应的化学计量点的电位相一致，以减小滴定终点的误差。氧化还原指示剂是氧化还原滴定的通用指示剂。选择的原则是指示剂的条件电位应处在滴定突跃范围内。

Table 5-2 Some important $E_{In}^{\ominus'}$ of redox indicators and color changes

表 5-2 一些重要氧化还原指示剂的 $E_{In}^{\ominus'}$ 及颜色变化

Indicators 指示剂	$E_{In}^{\ominus'}$ / V ([H$^+$]=1mol·L^{-1})	Color changes 颜色变化	
		Oxidation state 氧化态	Reduced state 还原态
Methylthionine chloride 亚甲基蓝	0.36	Blue 蓝	Achromaticity 无色
Diphenylamine 二苯胺	0.76	Purple 紫	Achromaticity 无色
Sodium diphenylamine sulfonate 二苯胺磺酸钠	0.84	Fuchsia 紫红	Achromaticity 无色
N-phenylanthranilic acid 邻苯氨基苯甲酸	0.89	Fuchsia 紫红	Achromaticity 无色
1,10-phenanthroline-ferrous complex ion 邻二氮菲-亚铁	1.06	Light blue 浅蓝	Red 红
Nitrophenanthroline-iron complex 硝基邻二氮菲-亚铁	0.25	Light blue 浅蓝	Fuchsia 紫红

Section 3 Permanganate Titration

第 3 节 高锰酸钾法

1. Introduction

1. 概述

KMnO$_4$ was used as titrant in this method. KMnO$_4$ is a strong oxidant, and its oxidizingability and reduction products are related to the acidity of solution. In a strong acid solution, KMnO$_4$ is reduced to Mn^{2+}:

本法以 KMnO$_4$ 作滴定剂。KMnO$_4$ 是一种强氧化剂，它的氧化能力和还原产物都与溶液的酸度有关。在强酸性溶液中，KMnO$_4$ 被还原为 Mn^{2+}。

$$MnO_4^- + 8H^+ + 5e \longrightarrow Mn^{2+} + 4H_2O \qquad E^{\ominus}=1.51V$$

In weakly acidic, neutral or weakly alkaline solution, KMnO$_4$ is reduced to MnO$_2$ solution.

在弱酸性、中性或弱碱性溶液中，KMnO$_4$ 被还原为 MnO$_2$。

$$MnO_4^- + 2H_2O + 3e \longrightarrow MnO_2 + 4OH^- \qquad E^{\ominus}=0.588V$$

In strong alkaline solution, MnO$_4^-$ is reduced to MnO$_4^{2-}$.

在强碱性溶液中，MnO$_4^-$ 被还原成 MnO$_4^{2-}$。

$$MnO_4^- + e \longrightarrow MnO_4^{2-} \qquad E^{\ominus}=0.564V$$

KMnO$_4$ has stronger oxidation ability in strong acid solution, it also produces colorless Mn^{2+}, that is convenient for observation of titration end point. So, it is generally

由于 KMnO$_4$ 在强酸性溶液中有更强的氧化能力，同时生成无色的 Mn^{2+}，便于滴定终点的观察，因此一般都在强

used under strong acidic conditions. However, under alkaline conditions, reaction rate of KMnO₄ to oxidize organics is faster than that of acid. Therefore, when potassium permanganate is used to determine organics, most of them are used in alkaline solution (greater than 2 mol·L⁻¹ NaOH solution).

Potassium permanganate can be used to titrate directly many reducing substances, such as Fe^{2+}, As(Ⅲ), Sb(Ⅲ), W(Ⅴ), H_2O_2, $C_2O_4^{2-}$, NO_2^- and other reducing substances (including many organic substances) etc. Back titration can be used to determine some oxidizing substances MnO_2, PbO_2, etc.; some non-redox substances such as Ca^{2+}, Th^{4+} and rare earth ions can also be determined indirectly through reaction of MnO_4^- and $C_2O_4^{2-}$.

The advantage of potassium permanganate is its strong oxidizing ability, which can directly or indirectly determine many inorganic and organic substances. It can be used as an indicator in titration. But its disadvantage is that standard solution is not stable, reaction process is more complex, easily occurs side reactions and has poor selectivity. However, if the standard titration solution is properly prepared and stored, and the conditions are strictly controlled during the titration, most of these shortcomings can be overcome.

2. Preparation of KMnO₄ Standard Titration Solution

(1) Preparation Because potassium permanganate reagent often contains a small amount of MnO_2 and other impurities, distilled water used also often contains a small amount of reducing substances such as dust, organic matter, etc. These substances can reduce KMnO₄, so KMnO₄ standard solution cannot be directly prepared. Usually solution of similar concentration is prepared before calibration. When preparing, at first we weigh slightly more than theoretical amount of KMnO₄, dissolve it in a certain volume of distilled water, slowly boil it for 15min, cool it, leave it in the dark place for two weeks, and filter it with a treated No. 4 glass colander. At last, store it in brown reagent bottle.

酸性条件下使用。但是，在碱性条件下 KMnO₄ 氧化有机物的反应速率比在酸性条件下更快，所以用高锰酸钾法测定有机物时，大都在碱性溶液中（大于 2mol·L⁻¹ 的 NaOH 溶液）进行。

应用高锰酸钾法，可直接滴定许多还原性物质，如 Fe^{2+}、As(Ⅲ)、Sb(Ⅲ)、W(Ⅴ)、H_2O_2、$C_2O_4^{2-}$、NO_2^- 以及其他还原性物质（包括很多有机物）等；采用返滴定法可以测定某些具有氧化性的物质如 MnO_2、PbO_2 等；还可以通过 MnO_4^- 与 $C_2O_4^{2-}$ 的反应间接测定一些非氧化还原物质，如 Ca^{2+}、Th^{4+} 和稀土金属离子等。

高锰酸钾法的优点是氧化能力强，可直接或间接地测定许多无机物和有机物，在滴定时自身可作指示剂；其缺点是标准溶液不太稳定，反应历程比较复杂，易发生副反应，滴定的选择性也较差。但若标准滴定溶液配制、保存得当，滴定时严格控制条件，这些缺点大多可以克服。

2. KMnO₄ 标准滴定溶液的制备

（1）配制 因为高锰酸钾试剂中常含有少量的 MnO_2 和其他杂质，使用的蒸馏水中也常含有少量如尘埃、有机物等还原性物质，这些物质都能使 KMnO₄ 还原，所以 KMnO₄ 标准溶液不能直接配制，通常先配制成近似浓度的溶液后再进行标定。配制时，首先称取略多于理论用量的 KMnO₄，溶于一定体积的蒸馏水中，缓缓煮沸 15min，冷却，于暗处放置两周，用已处理过的 4 号玻璃滤埚过滤，贮于棕色试剂瓶中。

Preparation of KMnO$_4$ solution
高锰酸钾溶液的制备

Calibration of KMnO$_4$ solution
高锰酸钾溶液的标定

(2) Calibration The standard substances for calibrating KMnO$_4$ solutions include Na$_2$C$_2$O$_4$、H$_2$C$_2$O$_4$ · 2H$_2$O、(NH$_4$)$_2$Fe(SO$_4$)$_2$ · 2H$_2$O、As$_2$O$_3$, and pure iron wire. Among them, Na$_2$C$_2$O$_4$ is the most commonly used, because it is easy to purify, has stable properties, and does not contain crystal water. Na$_2$C$_2$O$_4$ is dried at 105-110 ℃ to be constant weight and ready for use. In H$_2$SO$_4$ solution, formula of reaction between MnO$_4^-$ and C$_2$O$_4^{2-}$ is

$$2MnO_4^- + 5C_2O_4^{2-} + 16H^+ \longrightarrow 2Mn^{2+} + 10CO_2\uparrow + 8H_2O$$

In order to make reaction quantitative, pay attention to the following titration conditions.

① Temperature The reaction rate is extremely slow at room temperature and needs to be heated to 65℃ for titration. If the temperature exceeds 90 ℃, H$_2$C$_2$O$_4$ will be partially decomposed, resulting in higher calibration results.

$$H_2C_2O_4 \longrightarrow H_2O + CO_2\uparrow + CO\uparrow$$

② Acidity If acidity is too low, MnO$_4^-$ will be partially reduced to MnO$_2$; if acidity is too high, decomposition of H$_2$C$_2$O$_4$ will be promoted. Generally, appropriate acidity at the beginning of the titration is about 1 mol · L^{-1}. To prevent reaction that induces the oxidation of Cl$^-$, titration in HCl medium should be avoided as much as possible, usually in H$_2$SO$_4$ medium.

③ Titration rate Reaction rate of MnO$_4^-$ and C$_2$O$_4^{2-}$ is very slow at the beginning, and reaction rate is gradually accelerated after Mn^{2+} formationed . Therefore, the second drop should be added after the first drop of KMnO$_4$ solution fades. With the progress of the titration, titration rate can be appropriately accelerated. However, it should not be too fast, otherwise KMnO$_4$ dropped into it will have no time to react with C$_2$O$_4^{2-}$, and it will decompose in the hot acid solution, resulting in low calibration result.

（2）标定 标定 KMnO$_4$ 溶液的基准物质有 Na$_2$C$_2$O$_4$、H$_2$C$_2$O$_4$ · 2H$_2$O、(NH$_4$)$_2$Fe(SO$_4$)$_2$ · 2H$_2$O、As$_2$O$_3$ 和纯铁丝等，其中最常用的是 Na$_2$C$_2$O$_4$，它易于提纯、性质稳定、不含结晶水。Na$_2$C$_2$O$_4$ 在 105～110℃烘干至恒重，即可使用。在 H$_2$SO$_4$ 溶液中，MnO$_4^-$ 与 C$_2$O$_4^{2-}$ 的反应为：

应注意以下条件。

① 温度 此反应在室温下速率极慢，需加热至 65℃ 滴定。若温度超过 90℃，则 H$_2$C$_2$O$_4$ 部分分解，导致标定结果偏高。

② 酸度 酸度过低，MnO$_4^-$ 会被部分地还原成 MnO$_2$；酸度过高，会促进 H$_2$C$_2$O$_4$ 分解。一般滴定开始时适宜酸度约为 1mol · L^{-1}。为防止诱导氧化 Cl$^-$ 的反应发生，应当尽量避免于 HCl 介质中滴定，通常在 H$_2$SO$_4$ 介质中进行。

③ 滴定速率 MnO$_4^-$ 与 C$_2$O$_4^{2-}$ 的反应开始时速率很慢，当有 Mn^{2+} 生成之后反应速率逐渐加快。因此，应等加入第一滴 KMnO$_4$ 溶液褪色后再加第二滴，随着滴定的进行，滴定速率可适当加快。但不宜过快，否则滴入的 KMnO$_4$ 来不及和 C$_2$O$_4^{2-}$ 反应，就在热的酸性溶液中分解，导致标定结果偏低。

$$4MnO_4^- + 12H^+ \longrightarrow 4Mn^{2+} + 5O_2\uparrow + 6H_2O$$

If a small amount of MnSO₄ is added as a catalyst before titration, it can be carried out at a faster rate in the initial stage of the titration.

④ End point of titration   Titrate with KMnO₄ solution until solution shows pale pink color. It will not fade in 30 seconds, that is the end point. In the process of keeping solution static, reducing substances in the air can also reduce KMnO₄ and discolor. Calibrated KMnO₄ solution is left for a period of time, if it is found that MnO(OH)₂ precipitates, it should be filtered and calibrated again.

3. Application Example of Permanganate Titration

(1) Direct titration—determination of H_2O_2 In an acidic solution, H_2O_2 is oxidized quantitatively by MnO_4^-.

$$5H_2O_2 + 2MnO_4^- + 6H^+ \longrightarrow 2Mn^{2+} + 5O_2\uparrow + 8H_2O$$

The reaction can proceed smoothly at room temperature, and the reaction is slow at the beginning. The reaction is accelerated with formation of Mn^{2+}. A small amount of Mn^{2+} can also be added as a catalyst first. If H_2O_2 contains organic substances, the latter will also consume KMnO₄, which will make measurement result higher. At this time, iodometric method or cerium method should be used to determine H_2O_2 content. Peroxides of alkali metals and alkaline earth metals can be measured by the same method.

(2) Indirect titration—determination of Ca^{2+} Ca^{2+}, Th^{4+} etc. have no variable valence in solution, but based on the formation of oxalate precipitation, they can also be determined indirectly by potassium permanganate method. Take measurement of Ca^{2+} as an example, at first it is precipitated as CaC_2O_4, then filtered, washed and dissolved in a hot dilute H_2SO_4 solution, and finally titrated $H_2C_2O_4$ with a KMnO₄ standard solution. According to the amount of KMnO₄ consumed, Ca^{2+} content can be calculated indirectly. The relevant reaction formula is as follows:

若滴定前加入少量 MnSO₄ 为催化剂，则在滴定的最初阶段就可以较快的速率进行。

④ 滴定终点   用 KMnO₄ 溶液滴定至溶液呈现淡粉红色，30s 不褪色即为终点。溶液在放置过程中，空气中的还原性物质也能使 KMnO₄ 还原而褪色。标定好的 KMnO₄ 溶液在放置一段时间后，如果发现有 MnO(OH)₂ 沉淀析出，应重新过滤并标定。

3. 高锰酸钾法应用示例

（1）直接滴定法——H_2O_2 的测定   在酸性溶液中，H_2O_2 被 MnO_4^- 定量氧化。

此反应在室温下即可顺利进行，开始时反应较慢，随着 Mn^{2+} 生成而加速反应，也可以先加入少量 Mn^{2+} 作催化剂。若 H_2O_2 中含有机物质，后者也会消耗 KMnO₄，会使测定结果偏高，此时应改用碘量法或铈量法测定 H_2O_2 含量。碱金属及碱土金属的过氧化物，可采用同样的方法进行测定含量。

（2）间接滴定法 ——Ca^{2+} 的测定 Ca^{2+}、Th^{4+} 等在溶液中没有可变价态，但基于生成草酸盐沉淀，也可用高锰酸钾法间接测定。以 Ca^{2+} 的测定为例，先沉淀为 CaC_2O_4，再经过滤，洗涤后将沉淀溶于热的稀 H_2SO_4 溶液中，最后用 KMnO₄ 标准溶液滴定 $H_2C_2O_4$。根据所消耗的 KMnO₄ 的量，间接求得 Ca^{2+} 的含量。

Determination of H_2O_2
双氧水中 H_2O_2 含量的测定

Determination of Ca^{2+}
氧化还原法测定总钙

$$Ca^{2+} + C_2O_4^{2-} \longrightarrow CaC_2O_4 \downarrow$$

$$CaC_2O_4 + 2H^+ \longrightarrow Ca^{2+} + H_2C_2O_4$$

$$5H_2C_2O_4 + 2MnO_4^- + 6H^+ \longrightarrow 2Mn^+ + 10CO_2 \uparrow + 8H_2O$$

In order to ensure that quantitative reaction between Ca^{2+} and $C_2O_4^{2-}$ is complete, and to obtain CaC_2O_4 with larger particles, which are convenient for filtration and washing, corresponding measures must be taken as follows:

① Add excessive $(NH_4)_2C_2O_4$ to the acidic solution, and then slowly neutralize solution with dilute ammonia water until the methyl orange is yellow (pH 3.5 to 4.5), so that precipitate is slowly formed;

② It must be aged for a period of time after the precipitation is complete;

③ $C_2O_4^{2-}$ adsorbed on the surface of precipitate, was washed with distilled water. In order to reduce the loss of precipitation dissolution, the precipitate should be washed with as little cold water as possible until the solution contains no $C_2O_4^{2-}$.

If the precipitate happens in a neutral or weakly alkaline solution, some $Ca(OH)_2$ or basic calcium oxalate will be formed, which will make measurement result lower.

(3) Determination of MnO_2 by back titration Some substances that cannot be titrated directly with $KMnO_4$ solution, such as MnO_2, PbO_2, etc., can be determined by back titration. For example, the content of MnO_2 in Pyrite is determined by reaction of MnO_2 and $C_2O_4^{2-}$ in an acidic solution.

为了保证 Ca^{2+} 与 $C_2O_4^{2-}$ 之间能定量反应完全，并获得颗粒较大的 CaC_2O_4 沉淀，便于过滤洗涤，必须采取相应的措施：

① 在酸性试液中先加入过量 $(NH_4)_2C_2O_4$ 然后用稀氨水慢慢中和试液至甲基橙显黄色（pH 为 3.5～4.5），以使沉淀缓慢地生成；

② 沉淀完全后须放置陈化一段时间；

③ 用蒸馏水洗去沉淀表面吸附的 $C_2O_4^{2-}$，为减少沉淀溶解损失，应当用尽可能少的冷水洗涤沉淀，洗至洗涤液中不含 $C_2O_4^{2-}$ 为止。

若在中性或弱碱性溶液中沉淀，会有部分 $Ca(OH)_2$ 或碱式草酸钙生成，将使测定结果偏低。

（3）返滴定法 MnO_2 的测定 一些不能直接用 $KMnO_4$ 溶液滴定的物质，如 MnO_2、PbO_2 等，可以用返滴定法测定。例如，软锰矿中 MnO_2 含量的测定是利用 MnO_2 和 $C_2O_4^{2-}$ 在酸性溶液中的反应。

$$MnO_2 + C_2O_4^{2-} + 4H^+ \longrightarrow Mn^{2+} + 2CO_2 \uparrow + 2H_2O$$

Add a certain amount of excessed $Na_2C_2O_4$ to the levigated ore sample, add H_2SO_4 and heat (the temperature must not be too high, otherwise $Na_2C_2O_4$ will be decomposed, affecting the accuracy of measurement results), when there are no brown black particles in the sample, it means the sample is completely decomposed. Then back-titrate the remaining oxalic acid with $KMnO_4$ standard solution while hot.

加入一定量过量的 $Na_2C_2O_4$ 于磨细的矿样中，加 H_2SO_4 并加热（温度不能过高，否则将使 $Na_2C_2O_4$ 分解，影响测定结果的准确度），当试样中无棕黑色颗粒存在时，表示试样分解完全。然后用 $KMnO_4$ 标准溶液趁热返滴定剩余的草酸。

$$2MnO_4^- + 5C_2O_4^{2-} + 16H^+ \longrightarrow 10CO_2 \uparrow + 2Mn^{2+} + 8H_2O$$

According to the difference between amount of $Na_2C_2O_4$ added and the amount of $KMnO_4$ consumed, the content of MnO_2 was calculated.

(4) Determination of organics   In strong alkaline solution, excess $KMnO_4$ can oxidize certain organic substances quantitatively. It is itself reduced to green MnO_4^{2-}. Based on reaction, these organics can be determined by potassium permanganate method. Taking measurement of formic acid as an example, if a certain amount of $KMnO_4$ standard titration solution is added to a strongly alkaline sample containing formic acid, the following reactions will occur.

$$HCOO^- + 2MnO_4^- + 3OH^- \longrightarrow CO_3^{2-} + 2MnO_4^{2-} + 2H_2O$$

After reaction is completed, the solution is acidified, MnO_4^{2-} is disproportionated to MnO_4^- and MnO_2, a certain amount of excessed $FeSO_4$ standard solution is added accurately, all high-valent manganese ions are reduced to Mn^{2+}, and the remaining $FeSO_4$ can be titrated with $KMnO_4$ standard solution. The content of formic acid was calculated from amount of $KMnO_4$ added twice and amount of $FeSO_4$. This method can also be used for determination of methanol, formaldehyde, glycerol, glycolic acid (Carboxylic acid), tartaric acid, citric acid, phenol, salicylic acid, glucose, etc.

(5) Determination of chemical oxygen demand (COD_{Mn})   COD is a comprehensive indicator that measures degree of pollution of water by reducing substances (mainly organic substances). It refers to the amount of oxidant consumed when reducing substances in 1L of water are oxidized under certain conditions, and is converted into the mass concentration of oxygen (mol · L^{-1}), COD has become one of the main items of environmental monitoring and analysis.

When measuring COD_{Mn}, H_2SO_4 and a certain amount of excess $KMnO_4$ standard titration solution are added to water sample, heat it in a boiling water to oxidize the reducing substances therein, and reduce the remaining $KMnO_4$ with a certain amount of excess $Na_2C_2O_4$, and

根据 $Na_2C_2O_4$ 的加入量和 $KMnO_4$ 溶液消耗量之差，求出 MnO_2 的含量。

（4）有机物的测定   在强碱性溶液中，过量 $KMnO_4$ 能定量地氧化某些有机物，自身被还原为绿色的 MnO_4^{2-}，利用这一反应，可用高锰酸钾法测定这些有机物。以甲酸的测定为例，在含有甲酸的强碱性试样中加入一定量过量的 $KMnO_4$ 标准滴定溶液，则会发生：

待反应完成后，将溶液酸化，MnO_4^{2-} 歧化为 MnO_4^- 和 MnO_2，准确加入一定量且过量的 $FeSO_4$ 标准溶液，将所有高价锰离子全部还原为 Mn^{2+}，再用 $KMnO_4$ 标准溶液滴定剩余的 $FeSO_4$。由两次加入 $KMnO_4$ 的量及 $FeSO_4$ 的量可计算甲酸的含量。此法还可用于测定甲醇、甲醛、甘油、甘醇酸（羧基乙酸）、酒石酸、柠檬酸、苯酚、水杨酸、葡萄糖等。

（5）化学需氧量（COD_{Mn}）的测定   COD 是衡量水体受还原性物质（主要是有机物）污染程度的综合性指标，它是指在一定条件下，1L 水中还原性物质被氧化时所消耗的氧化剂的量，换算成氧的质量浓度（以 mol · L^{-1} 计）来表示，目前 COD 已成为环境监测分析的主要项目之一。

测定 COD_{Mn} 时，在水样中加入 H_2SO_4 及一定量且过量的 $KMnO_4$ 标准溶液，置于沸水浴中加热，使其中的还原性物质氧化，剩余的 $KMnO_4$ 用一定量过量的 $Na_2C_2O_4$ 还原，再以 $KMnO_4$

then Back-titrate standard solution with $KMnO_4$. This method is suitable for determination of COD in clean water samples such as surface water and drinking water. For determination of COD in industrial wastewater and domestic sewage, $K_2Cr_2O_7$ method should be used.

标准溶液返滴定。该法适用于地表水、饮用水等较为清洁水样COD的测定，对于工业废水和生活污水COD的测定，应采用$K_2Cr_2O_7$法。

Section 4 Dichromate Titration
第4节 重铬酸钾法

1. Method Outline

$K_2Cr_2O_7$ is a strong oxidant, in an acidic medium, $Cr_2O_7^{2-}$ is reduced to Cr^{3+}. $K_2Cr_2O_7$ has lower oxidation ability in acidic solution than $KMnO_4$ and is not as widely used as the potassium permanganate method. However, compared with potassium permanganate method, potassium dichromate method has the following advantages.

1. 方法概要

$K_2Cr_2O_7$是一种强氧化剂，在酸性介质中，$Cr_2O_7^{2-}$被还原为Cr^{3+}。$K_2Cr_2O_7$在酸性溶液中的氧化能力不如$KMnO_4$强，应用范围不如高锰酸钾法广泛，但与高锰酸钾法相比，重铬酸钾法有如下优点：

$$Cr_2O_7^{2-} + 14H^+ + 6e \longrightarrow 2Cr^{3+} + 7H_2O$$

$$E^{\ominus}_{Cr_2O_7^{2-}/Cr^{3+}} = 1.33V$$

① $K_2Cr_2O_7$ is easy to purify. After drying to constant weight at 120 ℃, it can be directly weighed to prepare a standard solution.

② $K_2Cr_2O_7$ solution is stable and stored in a closed container, and its concentration can be unchanged for a long time.

③ $K_2Cr_2O_7$ is weaker in oxidation than $KMnO_4$ and has stronger selectivity than $KMnO_4$. At room temperature, when HCl concentration is lower than 3 mol·L^{-1}, $Cr_2O_7^{2-}$ does not induce oxidation of Cl^-, so titration can be performed in hydrochloric acid medium.

The reduction product Cr^{3+} of $Cr_2O_7^{2-}$ is green, and excessed $Cr_2O_7^{2-}$ which shows yellow cannot be discerned at the end point. Therefore, indicator must be added to indicate the end point. Commonly-used indicator is sodium diphenylamine sulfonate.

① $K_2Cr_2O_7$易于提纯，120℃干燥至恒重后，可直接称量配制标准溶液。

② $K_2Cr_2O_7$溶液稳定，保存在密闭容器中，其浓度可长期不变。

③ $K_2Cr_2O_7$氧化性较$KMnO_4$弱，选择性较$KMnO_4$强，室温下，当HCl浓度低于3mol·L^{-1}时，$Cr_2O_7^{2-}$不会诱导氧化Cl^-，因此滴定可在盐酸介质中进行。

$Cr_2O_7^{2-}$的还原产物Cr^{3+}呈绿色，终点时无法辨别出过量的$Cr_2O_7^{2-}$的黄色，因而须加入指示剂指示终点，常用的指示剂是二苯胺磺酸钠。

2. Application Example of Dichromate Titration

(1) Determination of total iron content in iron ore
Dichromate titration is a standard method for determination of total iron in iron ore. According to different pre-oxidation reduction methods, it is divided into $SnCl_2$-$HgCl_2$ method and $SnCl_2$-$TiCl_3$ method (mercury-free determination method).

① The $SnCl_2$-$HgCl_2$ method The sample is dissolved with hot concentrated hydrochloric acid and Fe^{3+} is reduced to Fe^{2+} with $SnCl_2$ while hot. After cooling, excess $SnCl_2$ is oxidized with $HgCl_2$, and then diluted with water, and H_2SO_4-H_3PO_4 mixed acid and Sodium diphenylamine sulfonate are added. Sodium indicator titrated with $K_2Cr_2O_7$ standard titration solution until the solution changes from light green (Cr^{3+} green) to purple red, which is the end point of titration, and its main reaction formula is as follows:

$$Fe_2O_3 + 6HCl \longrightarrow 2FeCl_3 + 3H_2O$$
$$2Fe^{3+} + Sn^{2+} \longrightarrow 2Fe^{2+} + Sn^{4+}$$
$$Sn^{2+} + 2HgCl_2 \longrightarrow Sn^{4+} + 2Cl^- + Hg_2Cl_2 \downarrow (白色)$$
$$6Fe^{2+} + Cr_2O_7^{2-} + 14H^+ \longrightarrow 6Fe^{3+} + 2Cr^{3+} + 7H_2O$$

The purpose of adding H_3PO_4 before titration is to make Fe^{3+} generate stable $Fe(HPO_4)_2^-$, reduce potential of Fe^{3+}/Fe^{2+} couples, increase jump range, and make potential of discoloration point of sodium diphenylamine sulfonate indicator fall in the range of titration, the end point error is reduced. At the same time, because $Fe(HPO_4)_2^-$ is colorless, yellow color of Fe^{3+} is eliminated, which is beneficial to observation of the end point.

Cu(Ⅱ)、Mo(Ⅵ)、As(Ⅴ)、Sb(Ⅴ) ions, these can be reduced by $SnCl_2$ and oxidized by $K_2Cr_2O_7$, which affects the determination of iron. If silicon content in the sample is high, it should be decomposed with HF-H_2SO_4 to eliminate interference of Si. If NO_3^- is present, H_2SO_4 should be added to heat to remove it. This method is simple, fast and accurate, which is widely used in production. However, the mercury salt used for pre-reduction is toxic

2. 重铬酸钾法应用示例

Determination of total iron content
重铬酸钾法测全铁

（1）铁矿石中全铁量的测定 重铬酸钾法是测定铁矿石中全铁量的标准方法，根据预氧化还原方法的不同，分为$SnCl_2$-$HgCl_2$法和$SnCl_2$-$TiCl_3$法（无汞测定法）。

① $SnCl_2$-$HgCl_2$法 试样用热浓盐酸溶解，用$SnCl_2$趁热将Fe^{3+}还原为Fe^{2+}，冷却后，过量的$SnCl_2$用$HgCl_2$氧化，再用水稀释，并加入H_2SO_4-H_3PO_4混合酸和二苯胺磺酸钠指示剂，用$K_2Cr_2O_7$标准滴定溶液滴定至溶液由浅绿色（Cr^{3+}绿色）变为紫红色，即为滴定终点。

在滴定前加入H_3PO_4的目的是使Fe^{3+}生成稳定的$Fe(HPO_4)_2^-$，降低Fe^{3+}/Fe^{2+}电对的电位，增大突跃范围，使二苯胺磺酸钠指示剂变色点的电位落在滴定突跃范围内，减小终点误差；同时，由于$Fe(HPO_4)_2^-$是无色的，消除了Fe^{3+}的黄色，有利于终点的观察。

Cu(Ⅱ)、Mo(Ⅵ)、As(Ⅴ)、Sb(Ⅴ)等离子存在，既能被$SnCl_2$还原，又会被$K_2Cr_2O_7$氧化，影响铁的测定。若试样中硅含量高时，宜用HF-H_2SO_4分解，以消除硅的干扰。如有NO_3^-存在，则应加入H_2SO_4加热以除去。此法简便、快速又准确，生产上广泛使用。但因预还原用的汞盐有毒，可引起环境污染，近年来出现了无汞测铁法。

and causes environmental pollution. In recent years, a mercury-free iron measurement method has appeared.

② $SnCl_2$-$TiCl_3$ method   After the sample is dissolved with acid, the majority of Fe^{3+} is reduced to Fe^{2+} with $SnCl_2$ while hot, and then Na_2WO_4 is used as an indicator, and $TiCl_3$ is dropped to reduce the remaining Fe^{3+}.

$$2Fe^{3+} + Sn^{2+} \longrightarrow 2Fe^{2+} + Sn^{4+}$$

$$Fe^{3+} + Ti^{3+} \longrightarrow Fe^{2+} + Ti^{4+}$$

When Fe^{3+} is quantitatively reduced to Fe^{2+}, a slight excess of $TiCl_3$ will reduce W(Ⅵ) to W(Ⅴ), the latter is commonly known as Tungsten Blue, at this time the solution appears blue. After dilution with water, add $K_2Cr_2O_7$ solution until the blue just fades away, or use Cu^{2+} as a catalyst to oxidize a slight excess of $TiCl_3$ and tungsten blue by dissolved oxygen in water to fade the blue. The subsequent titration steps are the same as $SnCl_2$-$HgCl_2$ method.

It must be noted that if $SnCl_2$ is excessive, the measurement result will be higher, and excessive $TiCl_3$ will often cause precipitation of tetravalent titanium salts when diluted with water, which will affect the determination. When $TiCl_3$ is used to reduce Fe^{3+}, one drop of $TiCl_3$ can be added after the solution shows blue, otherwise Tungsten Blue fades too slowly. Add catalyst $CuSO_4$, and the titration must be done after Tungsten Blue fades for 1 minute. Because the slight excess Ti^{3+} is not cleaned, it is necessary to consume more $K_2Cr_2O_7$ standard titration solution to make the measurement result higher. Other oxidizing or reducing substances can also be determined through the reaction of $Cr_2O_7^{2-}$ and Fe^{2+}. For example, determination of chromium in steel, first oxidize chromium to $Cr_2O_7^{2-}$ with appropriate oxidant, and then titrate with Fe^{2+} standard solution.

(2) Determination of chemical oxygen demand (COD_{Cr})   Potassium dichromate is used as an oxidant in acidic medium, and the measured chemical oxygen demand is recorded as COD_{Cr}. It is the most widely-used COD measurement method. The measurement steps are as follows: add excessive $K_2Cr_2O_7$ standard titration

solution to the water sample. Ag_2SO_4 is used as a catalyst in acidic medium (H_2SO_4), and the mixture is refluxed for 2 hours, so that $K_2Cr_2O_7$ can fully oxidize organic and other reducing substances in the wastewater. After oxidation is complete, 1,10-phenanthroline-ferrous complex ion is used as an indicator, and Fe^{2+} standard solution titrates the remaining $K_2Cr_2O_7$. If the Cl^- content in the water sample is high, added $HgSO_4$ to eliminate the interference. The method has a wide range of applications and can be used for severely-polluted domestic sewage and industrial wastewater. The disadvantage is that $Cr(VI)$, Hg^{2+} and other harmful substances are used in the determination process, and the waste liquid needs to be treated.

（H_2SO_4）中，以 Ag_2SO_4 为催化剂，加热回流 2h，使 $K_2Cr_2O_7$ 充分氧化废水中有机物和其他还原性物质，待氧化作用完全后，以邻二氮菲-亚铁为指示剂，用 Fe^{2+} 标准滴定溶液滴定剩余的 $K_2Cr_2O_7$。如果水样中 Cl^- 含量高，需加入 $HgSO_4$ 以消除其干扰。该法适用范围广泛，可用于污染严重的生活污水和工业废水，缺点是测定过程中使用 $Cr(VI)$、Hg^{2+} 等有害物质，废液需处理。

Determination of chemical oxygen demand
重铬酸钾法测定 COD

Section 5　Iodometric Method
第 5 节　碘量法

The iodometric method is a titration method using oxidizing of I_2 and reducing of I^-. Because solubility of solid I_2 in water is very small (0.0013mol·L^{-1}) and easily volatile, I_2 is dissolved in KI solution. At this time, I_2 is present in the solution in the form of I_3^-:

碘量法是利用 I_2 的氧化性和 I^- 的还原性进行滴定的方法。由于固体 I_2 在水中的溶解度很小（0.0013mol·L^{-1}）且易挥发，所以将 I_2 溶解在 KI 溶液中，这时 I_2 以 I_3^- 形式存在溶液中。为方便

$$I_2 + I^- \rightleftharpoons I_3^-$$

In order to easily clarify stoichiometric relationship, it is generally abbreviated as I_2, and its half reaction formula is as follows

和明确化学计量关系，一般仍简写为 I_2，其半反应式：

$$I_2 + 2e \longrightarrow 2I^- \qquad E^{\ominus}=+0.5345V$$

From the value of electrode potential of the pair, it is known that I_2 is a weaker oxidant and can interact with a strong reducing agent. However, I^- is a medium-strength reducing agent that can interact with many oxidants. Therefore, a quantitative determination method is available in direct and indirect ways.

由电对的电极电位的数值可知，I_2 是较弱的氧化剂，可与较强的还原剂作用；而 I^- 则是中等强度的还原剂，能与许多氧化剂作用。因此，碘量法测定可用直接和间接的两种方式进行。

1. Direct Iodometric Method

Direct iodometry, also called iodometric titration, is a method of titrating directly some reducing substances by using I_2 standard solution. Reducing substances, whose electrode potentials is lower than $E^{\ominus}_{I_2/I^-}$, can be titrated directly with a standard solution of I_2. For example, for determination of sulfur in steel, the sample is burned with O_2 in a tube furnace at 1300 ℃, so that sulfur in the steel is converted to SO_2, and SO_2 is absorbed by water, and then titrated with the I_2 standard solution. Its reaction is:

$$I_2 + SO_2 + 2H_2O \longrightarrow 2I^- + SO_4^{2-} + 4H^+$$

The end point is very obvious by using starch as an indicator. Direct iodometry can be used to determine SO_2, S^{2-}, As_2O_3, $S_2O_3^{2-}$, Sn(Ⅱ), Sb(Ⅲ) and vitamin C, strongly reducing substances.

Direct iodometry cannot be performed in alkaline solution, otherwise I_2 will have a disproportionation reaction.

$$3I_2 + 6OH^- \longrightarrow IO_3^- + 5I^- + 3H_2O$$

Standard electrode potential of iodine is not high, direct iodometric method is not as widely used as indirect iodometric method.

2. Karl-Fischer Method for Determination of Moisture Content

The basic principle of this method is that quantitative H_2O is required when I_2 oxidizes SO_2:

$$I_2 + SO_2 + 2H_2O \rightleftharpoons H_2SO_4 + 2HI$$

The reaction is reversible, and it needs to be done in an alkaline solution. Generally, pyridine (C_5H_5N) is used as a solvent to make the reaction quantitatively to the right, its total reaction is:

1. 直接碘量法

直接碘量法又称碘滴定法，是利用 I_2 标准溶液直接滴定一些还原性物质的方法。电极电位比 $E^{\ominus}_{I_2/I^-}$ 小的还原性物质，可以直接用 I_2 的标准溶液滴定。例如，测定钢铁中硫，试样在 1300 ℃ 的管式炉中通 O_2 燃烧，使钢铁中的硫转化为 SO_2，用水吸收 SO_2，再用 I_2 标准溶液滴定。

Determination of vitamin C
直接碘量法测维生素 C 的含量

采用淀粉作指示剂，终点非常明显。用直接碘量法可以测定 SO_2、S^{2-}、As_2O_3、$S_2O_3^{2-}$、Sn(Ⅱ)、Sb(Ⅲ) 和维生素 C 等强还原性物质。

直接碘量法不能在碱性溶液中进行，否则 I_2 会发生歧化反应。

由于碘的标准电极电位不高，所以直接碘量法不如间接碘量法应用广泛。

2. 卡尔－费休法测定水含量

该方法的基本原理是当 I_2 氧化 SO_2 时需要定量的 H_2O。

这个反应是可逆的，反应需在碱性溶液中进行。一般采用吡啶（C_5H_5N）作溶剂，使反应定量地向右进行。

$$C_5H_5N \cdot I_2 + C_5H_5N \cdot SO_2 + C_5H_5N + H_2O \longrightarrow 2C_5H_5N \cdot HI + C_5H_5N \cdot SO_3$$

The generated $C_5H_5N \cdot SO_3$ can also react with water and consume some water, which interfers with the measurement, so methanol is added to prevent the above reaction :

生成的 $C_5H_5N \cdot SO_3$ 也能与水反应，消耗一部分水，因而干扰测定。为此加入甲醇，以防止上述反应发生。

$$C_5H_5N \cdot SO_3 + CH_3OH \longrightarrow C_5H_5NHOSO_2OCH_3$$

From the above discussion, it can be known that standard titration solution for determination of water by Karl-Fisher method is a mixed solution of I_2, SO_2, C_5H_5N, and CH_3OH, which is called a waste reagent. This reagent is reddish-brown in I_2 and light yellow after reaction with water. When the solution changes from light yellow to reddish brown, it is the end point. The utensils used in the measurement must be dry, otherwise it will cause errors. Reagent calibration can use water-methanol standard solution, or a stable crystalline hydrate as a reference substance.

由以上讨论可知，卡尔 - 费休法测定水的标准滴定溶液是 I_2、SO_2、C_5H_5N 和 CH_3OH 的混合溶液，称为费休试剂。此试剂呈 I_2 的红棕色，与水反应后呈浅黄色，当溶液由浅黄色变成红棕色即为终点，测定中所用器皿都须干燥，否则会造成误差。试剂的标定可用水 - 甲醇标准溶液，或以稳定的结晶水合物为基准物质。

This method is not only widely used to determine the moisture content of inorganic and organic substances, but also can be used to indirectly determine the content of various organic substances, such as alcohols, anhydrides, carboxylic acids, nitriles, and carbonyl groups, based on the amount of water produced or consumed in the reaction.

此法不仅广泛用于测定无机物和有机物中的水分含量，而且根据有关反应中生成水或消耗水的量，可以间接测定多种有机物的含量，如醇、酸酐、羧酸、腈类、羰基化合物、伯胺、仲胺以及过氧化物等。

Experiment 5 Determination of Hydrogen Peroxide

1. Objective

(1) Master the Principle and operation of determination of hydrogen peroxide with $KMnO_4$ standard solution.
(2) Master the sampling method of liquid sample.
(3) Master the preparation and preservation of a potassium permanganate solution.
(4) Master the principle, the condition and the method of standardization of $KMnO_4$ solution with the primary standard $Na_2C_2O_4$.
(5) Master the use of self-indicator in the detection of endpoint.

2. Principle

Potassium permanganate ($KMnO_4$) is a vigorous oxidant in an acidic solution, and its electrode reaction as follow,

$$MnO_4^- + 8H^+ + 5e \longrightarrow Mn^{2+} + 4H_2O$$

The titration reaction should be carried out in sulfuric acid (H_2SO_4), in general, the acidity of the solution should be maintained at $1 \sim 2$ mol $\cdot$ L^{-1}. $KMnO_4$ on sale usually contains impurities such as MnO_2, chloride, sulfate, nitrate and so on. Therefore, it cannot be used directly in the preparation of standard solution. Moreover, because the oxidization ability of $KMnO_4$ is strong, and it readily reacts with reductive substances such as organic impurities in water, ashes in air and so on, it easily decomposes when exposed to light. When it is prepared, its solution must be boiled or be dissolved with cold distilled water and then kept in brown reagent bottle in dark.

$KMnO_4$ solution generally can be standardized by sodium oxalate ($Na_2C_2O_4$). The reaction between $KMnO_4$ and $Na_2C_2O_4$ in an acidic solution is:

$$2MnO_4^- + 5C_2O_4^{2-} + 16H^+ \longrightarrow 2Mn^{2+} + 10CO_2 \uparrow + 8H_2O$$

Heating up is necessary because the reaction is slow. Even if so, the reaction is still slow and the color of $KMnO_4$ can't fade rapidly at the beginning of the titration. So at the beginning of titration, the $KMnO_4$ solution must be titrated drop by drop for speed of reaction is very slow. But the reaction is accelerated once Mn^{2+}, catalyzer of the reaction, forms during the reaction, so titration speed may be increased slightly. Because $KMnO_4$ solution itself has color, the endpoint can be indicated by the color of $KMnO_4$.

In acidic solution, $KMnO_4$ can oxidize H_2O_2 and it can be reduced, and the reaction is:

$$H_2O_2 - 2e \longrightarrow O_2 \uparrow + 2H^+$$

$$2MnO_4^- + 5H_2O_2 + 6H^+ \longrightarrow 2Mn^{2+} + 5O_2 \uparrow + 8H_2O$$

At the beginning of the reaction, the speed is slow, and the solution fades difficultly. When a small amount of Mn^{2+} is produced, the reaction speeds up by the auto catalytic of Mn^{2+}.

3. Apparatus and Materials

(1) Apparatus   acidburette (25 mL), conical flask (250 mL), pipette (1 mL), beaker (500 mL), cylinder (10mL, 100 mL), brown bottle (500 mL), fine-porosity and sintered-glass funnel, electronic analytical balance.

(2) Materials $KMnO_4$(A. R.), H_2SO_4 solution (1 mol $\cdot$ L^{-1}), 3% (w/V) H_2O_2 solution, $Na_2C_2O_4$.

4. Procedures

(1) Preparation of 0.02 mol/L $KMnO_4$ standard solution   Dissolve about 1.6g $KMnO_4$ in 500 mL of boiled distilled water in a brown bottle. The solution is allowed to stand in dark place for $7 \sim 10$ days, filter through a fme-porosity, sintered-glass funnel and preserved in another brown bottle.

(2) Standardization of $KMnO_4$ solution   Weigh out accurately about 0.15g of primary standard $Na_2C_2O_4$ into 250 mL conical flask. Dissolve in 100 mL of water and 5mL H_2SO_4. The solution is then heated to 75-85℃ using water bath. During continuous shaking, titrate it with $KMnO_4$ solution drop after drop only when the pink of the first drop fades, then add to about 15 mL $KMnO_4$ solution from a burette in it. After the color of $KMnO_4$ fades, titration will be continued until a pale pink color is observed and persists for 0.5 min. The temperature of the solution at the endpoint should not be less than 55℃. The concentration of $KMnO_4$ can be calculated by the following formula:

$$c_{KMnO_4} = \frac{2 \times w_{Na_2C_2O_4}}{5 \times V_{KMnO_4} \times \dfrac{M_{Na_2C_2O_4}}{1000}} \qquad M_{Na_2C_2O_4} = 134.0 \text{g} \cdot \text{mol}^{-1}$$

(3) Determination of Hydrogen Peroxide   Pipet out 1.00 mL of 3% (*w/V*) H$_2$O$_2$ solution into a 250 mL conical flask with 20 mL distilled water. Add 20 mL of 1mol·L^{-1} H$_2$SO$_4$ solution, and titrate with 0.02 mol·L^{-1} KMnO$_4$ standard solutions, until a faint but distinct reddish color is obtained. From the titration value calculate the concentration of the H$_2$O$_2$ solution:

$$H_2O_2(\%) = \frac{(cV)_{KMnO_4} \times 5 \times M_{H_2O_2}/100}{2 \times V_{Sample}} \times 100 (\text{g} \cdot \text{mL}^{-1}) \qquad M_{H_2O_2} = 34.02 \text{g} \cdot \text{mol}^{-1}$$

5. Notes

(1) Because H$_2$O$_2$ readily decomposes when heated, the titration must be carried on under room temperature.

(2) Commercial KMnO$_4$ can't be used directly to prepare a standard solution due to the existence of MnO$_2$ as an impurity, which can accelerate the decomposition of KMnO$_4$. MnO$_2$ must be eliminated by filtration, but filter paper cannot be used.

(3) Distilled water generally contains some organic compounds, which can deoxidize KMnO$_4$, so the water must be boiled before use.

(4) Light may accelerate the decomposition of KMnO$_4$, so a KMnO$_4$ solution must be preserved in a brown bottle and kept in dark place for 7-10 days.

(5) The reaction is fairly slow, so the titration shouldn't be proceed too fast. The temperature of the solution at the endpoint should not be less than 55℃.

6. Questions

(1) What other methods can be used to determine the concentration of H$_2$O$_2$?

(2) When determine the concentration of H$_2$O$_2$ with KMnO$_4$ solution, Can HNO$_3$, HCl, and HAc be used to control acidity? Why?

(3) Can the solution be heated when the content of H$_2$O$_2$ solution is measured with KMnO$_4$ standard solution? Why?

实验 5　过氧化氢含量的测定

1. 实验目的

（1）掌握高锰酸钾溶液的配制与保存。

（2）掌握用草酸钠作基准物质标定高锰酸钾溶液的原理、条件和方法。

（3）掌握使用自身指示剂判断滴定终点。

（4）掌握用高锰酸钾标准溶液测定过氧化氢含量的原理和方法。

（5）掌握液体样品的取样方法。

2. 实验原理

在酸性溶液中高锰酸钾是一种强氧化剂，其反应方程式如下：

$$MnO_4^- + 8H^+ + 5e \longrightarrow Mn^{2+} + 4H_2O$$

滴定在硫酸中进行，酸度一般维持在 1～2mol/L，市售的高锰酸钾一般含有杂质，如二氧化锰、氯化钠、硫酸盐、硝酸等，因而不能直接配制标准溶液使用。另外高锰酸钾的氧化能力很强，很容易与水中的有机杂质、空气中的灰尘等还原物质反应，光照易分解。当配制高锰酸钾溶液时，必须煮沸或用煮沸并冷却的蒸馏水溶解，并保存在棕色的试剂瓶中。高锰酸钾溶液通常用草酸钠标定，反应方程式如下：

$$2MnO_4^- + 5C_2O_4^{2-} + 16H^+ \longrightarrow 2Mn^{2+} + 10CO_2\uparrow + 8H_2O$$

该反应速度很慢，必须在加热条件下进行。在开始滴定时，由于高锰酸钾的颜色不能很快褪去，故滴定开始时，高锰酸钾必须逐滴加入，产生的二价锰离子可作为催化剂。由于高锰酸钾自身有颜色，故可以通过高锰酸钾的颜色来指示终点。

在酸性溶液中，高锰酸钾能氧化过氧化氢：

$$H_2O_2 - 2e \longrightarrow O_2\uparrow + 2H^+$$

$$2MnO_4^- + 5H_2O_2 + 6H^+ \longrightarrow 2Mn^{2+} + 5O_2\uparrow + 8H_2O$$

反应开始时速率很慢，而且溶液不容易褪色。当有少量二价锰离子生成时，反应速率则会加快。

3. 仪器和材料

（1）仪器 酸式滴定管（25mL），锥形瓶（250mL），烧杯（500mL），量筒（10mL，100mL），棕色试剂瓶（500mL），多孔玻璃烧结漏斗，电子分析天平，移液管（1mL）。

（2）材料 $KMnO_4$（A.R.），$Na_2C_2O_4$（基准物质），H_2SO_4（1mol/L），3%（质量浓度）过氧化氢溶液。

4. 实验步骤

（1）0.02mol·L^{-1} 高锰酸钾标准溶液的配制 用500mL蒸馏水溶解1.6g高锰酸钾，于棕色瓶中避光放置7～10天，用多孔玻璃烧结漏斗过滤并保存在。

（2）高锰酸钾标准溶液的标定 准确称取0.15g草酸钠基准物质，置于250mL的锥形瓶中，用100mL蒸馏水和5mL硫酸溶解，水浴加热到75～85℃。用高锰酸钾溶液逐滴滴加（仅当前一滴粉红色褪色时再滴加）该溶液，加入大约15mL高锰酸钾溶液。在高锰酸钾褪色后，滴定继续进行，直至呈淡粉色，并持续0.5min，终点溶液的温度不应低于55℃。高锰酸钾的摩尔浓度计算公式如下：

$$c_{KMnO_4} = \frac{2 \times w_{Na_2C_2O_4}}{5 \times V_{KMnO_4} \times \dfrac{M_{Na_2C_2O_4}}{1000}}$$

（3）过氧化氢含量的测定 准确移取1.00mL 3%的过氧化氢溶液，置于含有20mL蒸馏水的 250mL 锥形瓶中,再加入 20mL 1mol·L⁻¹ 的硫酸溶液,然后用 0.02mol·L⁻¹ 高锰酸钾标准溶液滴定,直到终点显示浅红色。用滴定值计算过氧化氢含量:

$$H_2O_2(\%) = \frac{(cV)_{KMnO_4} \times 5 \times M_{H_2O_2}/100}{2 \times V_{Sample}} \times 100 (g \cdot mL^{-1})$$

5. 注意事项

(1) 过氧化氢受热易分解,因此滴定必须在室温下进行。

(2) 市售的高锰酸钾不能直接配制成标准溶液,因为二氧化锰作为杂质存在,会加速高锰酸钾分解,二氧化锰可以通过过滤除去,注意不能用滤纸过滤。

(3) 蒸馏水通常包含一些能还原高锰酸钾的杂质,所以蒸馏水在用之前需要煮沸。

(4) 光照会加速高锰酸钾的分解,所以高锰酸钾溶液必须保存在棕色试剂瓶内,并在暗处放置 7~10 天。

(5) 反应很慢,因此开始滴定时不宜过快;滴定终点时溶液的温度不能低于 55℃。

6. 思考

(1) 测定过氧化氢含量还有其他方法吗?

(2) 当用高锰酸钾标准溶液测定过氧化氢溶液含量时,能采用硝酸、盐酸或醋酸控制酸度吗?为什么?

(3) 采用高锰酸钾标准溶液测定过氧化氢的含量时,能否加热?为什么?

Exercises

5-1 True or False

(1) In the redox reaction, the higher the potential of the electrical pair, the stronger the reduction ability of the reduced state.

(2) The greater the difference between the conditional potentials of the two electric pairs, the greater the conditional equilibrium constant K of the redox reaction, and the more complete the reaction is.

(3) Redox reactions are done quantitatively, generally, $\lg K' \leqslant 6$, $E_1^\ominus - E_2^\ominus \geqslant 0.4V$, such redox reactions can be applied to titration analysis.

(4) The reaction rate increases with increasing temperature.

(5) Reactions that are catalyzed by the products themselves are called autocatalytic reactions.

(6) Induction and catalytic reactions differ little as usual.

(7) The standard $KMnO_4$ solution can be titrated with $FeSO_4$ solution without adding additional indicators.

(8) Starch is the exclusive indicator of iodine method.

(9) The medium required for $KMnO_4$ titration must be acidic.

(10) Permanganate titration has the advantage of strong oxidation ability, which can be used directly or indirectly to determine many inorganic and organic substances, and can be used as an indicator when titrating.

(11) $K_2Cr_2O_7$ is easy to purify, but can not be directly weighed to make a standard solution.

(12) Direct iodimetry is for the direct titration of some reducing substances using I_2 standard solution. Reductive substances with lower electrode potentials than $E^{\ominus}_{I_2/I^-}$ can be titrated directly with a standard solution of I_2.

(13) Direct iodimetry must be carried out in an alkaline solution for the reaction to be complete.

(14) In the Karl-Fischer method of measuring water, all the vessels used must be dry, otherwise it will cause errors.

(15) In the determination of iron by potassium dichromate method, the main purpose of adding sulfur-phosphoric acid is to maintain acidic conditions.

(16) In iodimetry, a starch indicator should be added at the beginning of titration, otherwise the end point will be delayed.

(17) In the determination of COD by potassium dichromate method, if the content of Cl^- in water sample is high, $HgSO_4$ should be added to eliminate its interference.

(18) $KMnO_4$ is stable and pure, and its standard solution can be prepared directly.

(19) Redox indicator is a general indicator for redox titration. The principle of selection is that the conditional potential of the indicator should be in the titration jump range.

(20) In redox titration, only the potentiometric method can determine the endpoint.

5-2 Electrode potential can be used to judge the properties of redox reaction, but it cannot be defined ( ).
A. redox reaction rate
B. redox reaction direction
C. redox capacity
D. redox completeness

5-3 The factor affecting the equilibrium constant of redox reactions is ( ).
A. reactant concentration
B. catalyst
C. the temperature
D. induced reaction

5-4 The reference reagent for calibrating $Na_2S_2O_3$ solution is ( ).
A. $Na_2C_2O_4$
B. $(NH_4)_2C_2O_4$
C. Fe
D. K_2CrO_7

5-5 Positions of stoichiometric points in redox titration is ( ).
A. right in the middle of the titration jump
B. toward the side with more electron gains and losses
C. the side with less electronic gain and loss
D. Uncertain

5-6 Potassium dichromate titration method is used for iron determination, adding H_3PO_4 is mainly to ( ).
A. prevent the precipitation

B. improve acidity

C. reduce the potential of Fe^{3+}/Fe^{2+} and increase the jump range

D. prevent oxidation of Fe^{2+}

5-7 If the indirect iodimetry is carried out in alkaline medium, the results will be affected due to the disproportionation reaction of ( ).

A. $S_2O_3^{2-}$ B. I^-
C. I_2 D. $S_2O_4^{2-}$

5-8 In acidic media, titration of oxalate solution with $KMnO_4$ solution should be ( ).

A. as fast as acid-base titration

B. always proceed slowly

C. that titration rate is slow at the beginning, then gradually accelerates, and slows down near the end point

D. fast, then slow down

5-9 The following redox titration indicators belong exclusively to ( ).

A. diphenylamine sulfonate B. methylene blue
C. starch solution D. potassium permanganate

5-10 Known: $Fe^{3+} + e \longrightarrow Fe^{2+}$, $E^{\ominus}=0.771V$; $Cu^{2+} + 2e \longrightarrow Cu$, $E^{\ominus}=0.337V$; $Fe^{2+}+2e \longrightarrow Fe$, $E^{\ominus}=-0.440V$; $Al^{3+}+3e \longrightarrow Al$, $E^{\ominus}=-1.622V$. The strongest reducing agent is ( ).

A. Al^{3+} B. Fe^{2+}
C. Fe D. Al

5-11 In potassium permanganate calibrated with $Na_2C_2O_4$, the fading is slow at the beginning, but the fading became faster later because ( ).

A. the temperature is too low

B. the temperature rises after the reaction

C. Mn^{2+} is catalytic

D. potassium permanganate concentration decreases

5-12 The following standard solutions are generally prepared by direct method: ( ).

A. $KMnO_4$ standard solution B. I_2 standard solution
C. $K_2Cr_2O_7$ standard solution D. $Na_2S_2O_3$ standard solution

5-13 If Fe^{2+}, Cl^-, and I^- coexist in a solution, and I^- is removed by oxidation without affecting Fe^{2+} and Cl^-, the reagent to be added is( ).

A. Cl_2 B. $KMnO_4$ C. $FeCl_3$ D. HCl

5-14 The iodimetry method is used to determine the content of $CuSO_4$, and excessive KI is added into the sample solution to ( ).

A. reduce Cu^{2+} to Cu^+ B. prevent the volatilization of I_2
C. form CuI precipitation with Cu^+ D. reduce $CuSO_4$ to elemental Cu

5-15 Which of the following indicators can be used to determine Fe^{2+} by $KMnO_4$ method? ( )

A. Self-indicator B. Sodium diphenylamine sulfonate
C. Chrome black T D. Methyl red-bromocresol green

5-16 Potassium permanganate is not generally used for ( ).
   A. direct titration   B. displacement titration
   C. backtitration   D. indirect titration

5-17 Which of the following statements about galvanic cells is false? ( ).
   A. The electrode into which electrons flow is the positive pole.
   B . The pole at which oxidation occurs is the positive pole.
   C. Electrons leave is the negative terminal.
   D. The less active metal pole is the positive pole.

5-18 Standard solution of I_2 is prepared by dissolving I_2 in ( ).
   A. water   B. KI solution   C. HCl solution   D. KOH solution

5-19 Which of the following factors is independent of the size of electrode potential? ( ).
   A. Properties of electrode itself
   B. Temperature
   C. Concentration of oxidation and reduction states
   D. The writing of chemical equations

5-20 Which of the following statements is true? ( ).
   A. The lower the potential of A electric pair, the stronger the oxidation capacity of its oxidation form
   B. The higher the potential , the stronger the oxidation capacity of the oxidized form
   C. The higher the potential of the electrical pair , the stronger the reduction ability of the reduced form
   D. An oxidant can oxidize a reducing agent with a higher potential than it

5-21 When the iodine value of vegetable oil is determined by indirect iodinometry, the indicator starch solution should be added ( ).
   A. before titration begins   B. halfway through titration
   C. at the end point of titration   D. after stoichiometric point

5-22 In 1mol·L^{-1} HCl solution, titrate Sn^{2+} with Fe^{3+}, and calculate the potential value when 99.9% of Sn^{2+} is titrated to produce Sn^{4+}. (Known $E^{\ominus}_{Fe^{3+}/Fe^{2+}}$ =0.70V, $E^{\ominus}_{Sn^{4+}/Sn^{2+}}$ =0.14V)

(0.23V)

5-23 A copper sheet is inserted into a beaker containing 0.5mol·L^{-1} $CuSO_4$ solution, and a silver sheet is inserted into a beaker containing 0.5mol·L^{-1} $AgNO_3$ solution, and they are connected to form a galvanic cell. [Known $E^{\ominus}_{Ag^+/Ag}$ =0.7996V ; $E^{\ominus}_{Cu^{2+}/Cu}$ =0.342V]
   (a) Write out the symbol of the galvanic cell;(b) write out the electrode reaction formula and the battery reaction of the galvanic cell;(c) Find the electromotive force of the battery.

(0.4665V)

5-24 Weigh the reference substance $Na_2C_2O_4$ 0.1500g dissolved in a strong acid solution, titrate with $KMnO_4$ standard solution, 20.00mL is used when reaching the end point, and calculate the molar concentration of $KMnO_4$ standard solution.

(0.02239 mol·L^{-1})

5-25 Dissolve 0.5000 g limestone sample in hydrochloric acid, the calcium is precipitated to CaC_2O_4. After washing, the precipitation is dissolved in dilute H_2SO_4 solution. Titration is carried out with 0.02000 mol·L^{-1} $KMnO_4$ standard solution 38.05 mL is consumed at the end point to calculate the content of $CaCO_3$ in limestone.

(38.09%)

Chapter 6 Gravimetric Analysis and Precipitation Titration
第 6 章 重量分析法和沉淀滴定法

 Study Guide 学习指南

Gravimetric analysis is one of the typical chemical analysis methods for the quantitative determination of an analyte based on the measurement of mass. It can be classified into three types: volatilization gravimetry, precipitation gravimetry and electrogravimetry. In this charpter, we shall focus on the former two types. The target of this charpter is to familiarize with the mechanism, analytical procedure and result calculation of the gravimetric analysis methods.

重量分析法是经典的化学分析方法之一,它是根据生成物的质量来确定被测组分含量的方法。通常有汽化法、沉淀法和电解法,本章重点介绍汽化法和沉淀法。通过本章的学习,应掌握重量分析法的原理、测定过程及结果计算。

Section 1 Introduction of Gravimetric Analysis
第 1 节 重量分析法概述

1. Classification of Gravimetric Analysis

1. 重量分析法的分类

Gravimetric analysis is to isolate the analyte from other compositions by a proper method and determine its mass by weighing. According to different isolation methods, gravimetric analysis is generalized into the following types:

重量分析法是用适当的方法先将试样中的待测组分与其他组分分离,然后用称量的方法测定该组分的含量。根据分离方法的不同,重量分析法常分为三类。

(1) Vaporization or volatilization gravimetry　Based on the volatility of substances, the analyte is volatized by heating or other means. Then, calculate the content of the analyte according to the mass loss of the sample.

（1）汽化法（又称挥发法）　利用物质的挥发性质,通过加热或其他方法使试样中的待测组分挥发逸出,然后根据试样质量的减少,计算该组分的含量;

Another way to determine the analyte is to absorb the vapor emitted and to calculate the mass gain of the absorbent.

(2) Precipitation gravimetry　Precipitation gravimetry is the major method of gravimetric analysis. It involves formation of sparingly soluble precipitate by the adding of a reagent, filtering, washing, drying or ignition, and finally weighing of the precipitate and calculation. For example, the amount of SO_4^{2-} can be determined by precipitating it with excess $BaCl_2$ solution to form sparingly soluble $BaSO_4$ precipitate. After the precipitate is filtered, washed, dried and heated, it is weighed to get the mass of SO_4^{2-} in the original sample.

(3) Electrogravimetry　The metal ion under analysis is separated by electrolysis and deposited on an electrode. The amount of the ion can be known by the weight increase of the electrode.

2. Characteristics of Gravimetric Analysis

Gravimetric analysis is a typical method in analytical chemistry. It provides result directly by weighing without the introduction of a set of data from analytical facilities or the comparison with standard specimen or primary standard substance. Gravimetric analysis allows high degree of accuracy with relative error less than 0.1% for the determination of high content analyte. Until now, the accurate analysis of high content silicon, phosphorus, tungsten, nickel and rare earth still uses gravimetric analysis method. However, this method is convoluted and time consuming, not suitable for the control analysis in the production. What's more, the analytical accuracy is low for the determination of low content analyte.

Section 2　Volatilization Gravimetry

Based on the volatility of substances, the analyte is volatized by heating or other means. Then, calculate the

或者用吸收剂吸收逸出的组分，根据吸收剂质量的增加计算该组分的含量。

（2）沉淀法　沉淀法是重量分析法中的主要方法，这种方法是利用试剂与待测组分生成溶解度很小的沉淀，经过滤、洗涤、烘干或灼烧成为组成一定的物质，然后称其质量，再计算待测组分的含量。例如，测定试样中 SO_4^{2-} 含量时，可在试液中加入过量 $BaCl_2$ 溶液，使 SO_4^{2-} 完全生成难溶的 $BaSO_4$ 沉淀，经过滤、洗涤、烘干、灼烧后，称量 $BaSO_4$ 的质量，再计算试样中的 SO_4^{2-} 的含量。

（3）电解法　利用电解的方法使待测金属离子在电极上还原析出，然后称量，根据电极增加的质量，求得其含量。

2. 重量分析法的特点

重量分析法是经典的化学分析法，它通过直接称量得到分析结果，不需要从容器器皿中引入许多数据，也不需要标准试样或基准物质作比较。对高含量组分的测定，重量分析法比较准确，一般测定的相对误差不大于0.1%。对高含量的硅、磷、钨、镍、稀土元素等试样的精确分析，至今仍常使用重量分析法。但重量分析法的不足之处是操作较繁琐，耗时多，不适于生产中的控制分析；对低含量组分的测定误差较大。

第2节　汽化法

汽化法利用物质的挥发性质，通过加热或其他方法使试样中的待测组分挥

content of the analyte according to the mass loss of the sample. Another way to determine the desired substance is to absorb the vapor emitted and to calculate the mass gain of the absorbent.

1. Direct Method

If several volatiles co-exist, a proper absorbent has to be chosen so that only the analyte is absorbed. Then the increased weight of the absorbent is got. In this method, the analyte is weighed directly, so it is called the direct method. Because high temperature oxidation is often used in the direct method, it is also called the ignition method.

The ash content determination of the purity of pharmaceuticals in pharmacopoeia also adopts the ignition method, for instance, residue on ignition, inorganic impurities in organics. Pharmacopoeia stipulates that residue on ignition shall not exceed a specified limit.

$$\text{residue on ignition} = \frac{\text{mass of residue}}{\text{mass of sample}} \times 100\% \tag{6.1}$$

1. 直接法

若有几种挥发性物质并存时，选择适当吸收剂，定量吸收待测组分而不吸收其他共存物，根据称量吸收剂增重测量待测物质的含量——这种称为直接法。由于直接法常用高温氧化，故也称为灼烧法。

《中国药典》中，药物纯度检查灰分测定即用灼烧法，如灼烧残渣、检查有机物是否含无机杂质。《中国药典》检查项目规定，灼烧残渣不能超过一定限度。

$$\text{灼烧残渣} = \frac{\text{残渣质量}}{\text{样品质量}} \times 100\%$$

2. Indirect Method

Analyte is determined according to the weight loss of sample such as weight loss on drying(LOD) including crystal water, hygroscopic water and volatilizable substances under the same conditions.

Example: glucose

$C_6H_{12}O_6 \cdot H_2O \longrightarrow C_6H_{12}O_6 + H_2O \uparrow$ till weight constant

2. 间接法

间接法是根据样品减少的质量加以测定的方法，如测量干燥失重，减少的质量包括结晶水、吸湿水及该条件下能挥发的物质。

$$\text{LOD} = \frac{S-W}{S} \times 100\% \tag{6.2}$$

S——sample weight;
W——weight after drying

Section 3　Precipitation Gravimetry

第3节　沉淀法

In precipitation gravimetry, the sample is dissolved first, then a proper precipitating reagent is added to let analyte component in a precipitation reaction and produce a precipitation form. The formed precipitate is filtered, washed, dried or ignited under certain temperature to be converted into a weighing form which is to be weighed. The amount of analyte in the original sample can then be calculated according to the formula of the weighing form. Precipitation form and weighing form may be same or different. For example:

利用沉淀重量法进行分析时，首先需将试样分解为试液，然后加入适当的沉淀剂使其与待测组分发生沉淀反应，并以沉淀形态沉淀出来。沉淀经过过滤、洗涤，在适当的温度下烘干或灼烧，转化为称量形态，再进行称量。根据称量形态的化学式计算待测组分在试样中的含量。沉淀形态和称量形态可能相同，也可能不同。

$$Ba^{2+} \xrightarrow{\text{precipitate}} BaSO_4 \xrightarrow{\text{firing}} BaSO_4$$

$$Fe^{3+} \xrightarrow{\text{precipitate}} Fe(OH)_3 \xrightarrow{\text{firing}} Fe_2O_3$$

| 待测组分 | 沉淀形态 | 称量形态 |
| Analyte | Precipitation Form | Weighing Form |

1. Requirements for Precipitation Form and Weighing Form in Precipitation Gravimetry

1. 沉淀法对沉淀形态和称量形态的要求

The following requirements have to be met by precipitation form and weighing form in order to obtain accurate analytical result.

在重量分析法中，为获得准确的分析结果，沉淀形态和称量形态必须满足以下要求。

(1) Requirements for precipitation form

（1）对沉淀形态的要求

① The analyte must be completely precipitated. The precipitate is of such low solubility that losses from dissolution should not exceed the weighing error of analytical balance, less than 0.1mg in general. For example, two precipitation forms $CaSO_4$ and CaC_2O_4 are produced when determining Ca^{2+}, in which $CaSO_4$ ($K_{sp}=2.45\times10^{-5}$) has higher solubility than CaC_2O_4 ($K_{sp}=1.78\times10^{-9}$). Obviously, $(NH_4)_2C_2O_4$ is a better precipitating reagent than sulfuric acid.

① 对于沉淀形态，要求沉淀要完全，沉淀的溶解度要小，测定过程中沉淀的溶解损失不应超过分析天平的称量误差，一般要求溶解损失应小于0.1mg。例如，测定Ca^{2+}时，以形成$CaSO_4$和CaC_2O_4两种沉淀形态作比较，$CaSO_4$的溶解度较大（$K_{sp}=2.45\times10^{-5}$），CaC_2O_4的溶解度小（$K_{sp}=1.78\times10^{-9}$）。显然，用（NH_4）$_2C_2O_4$作沉淀剂比用硫酸作沉淀剂沉淀得更完全。

② The analyte must be highly pure and easily filtered and washed. High purity of precipitate is one of the

② 沉淀必须纯净，并易于过滤和

key factors in getting accurate analytical result. The crystalline precipitate with larger particle size (e. g. $MgNH_4PO_4 \cdot 6H_2O$) presents smaller surface area which would have slight chance to be contaminated. Therefore, it is pure and easily filtered and washed. On the contrary, the crystalline precipitate with smaller particle size (e. g. CaC_2O_4, $BaSO_4$) has larger surface area due to certain reasons, onto which impurities are adsorbed and washing times have to be prolonged accordingly. Non-crystalline precipitate has large and loose volume, attracting a large amount of impurities which take time to be filtered and washed. For such precipitates, a suitable precipitating condition has to be created to meet requirements for precipitation forms.

③ The precipitation form should be easily converted into weighing form through drying or ignition. For example, if analyte Al^{3+} is precipitated into $Al(C_9H_6NO)_3$, it can be dried at 130 ℃ and weighed. However, if it is precipitated into $Al(OH)_3$, it has to be ignited at 1200 ℃ in order to be converted into the oxide Al_2O_3 before being weighed. Therefore, the former method is better than the latter.

(2) Requirements for weighing form

① The composition of the weighing form must conform to its empirical formula. This is the basis for quantitative measurement. For instance, PO_4^{3-} can be precipitated into ammonium phosphomolybdate, but this precipitate cannot be used as weighing form to measure PO_4^{3-} due to its inconstancy in chemical composition. However, if quinoline phosphomolybdate gravimetric method is applied, a weighing form with composition corresponding to formula can be produced.

② The weighing form has to be sufficiently stable, free of influence from CO_2 and H_2O in the air. For instance, calcium ion might be precipitated to calcium oxalate $CaC_2O_4 \cdot H_2O$. It might then be heated to convert it into the oxide CaO. CaO tends to absorb H_2O and CO_2 in the air. Hence CaO is not suitable as a weighing form.

③ The molar mass of weighing form should be as large as possible to increase the mass of weighing

洗涤。沉淀纯净是获得准确分析结果的重要因素之一。颗粒较大的晶体沉淀（如 $MgNH_4PO_4 \cdot 6H_2O$）表面积较小，吸附杂质的机会较少，因此沉淀较纯净，易于过滤和洗涤。颗粒细小的晶形沉淀（如 CaC_2O_4、$BaSO_4$）比表面积大，吸附杂质多，洗涤次数也相应增多。非晶形沉淀（如 $Al(OH)_3$、$Fe(OH)_3$）体积庞大疏松、吸附杂质较多，过滤费时且不易洗净，对于这类沉淀，必须选择适当的沉淀条件以满足对沉淀形态的要求。

③ 沉淀形态经烘干、灼烧时，应易于转化为称量形态。例如，对于 Al^{3+} 的测定，若沉淀为 8-羟基喹啉铝 $[Al(C_9H_6NO)_3]$，在 130 ℃ 烘干后即可称量；而沉淀为 $Al(OH)_3$ 时，则必须在 1200 ℃ 灼烧才能转变为无吸湿性的 Al_2O_3，方可称量。因此，测定 Al^{3+} 时选用前法比后法好。

（2）对称量形态的要求

① 称量形态的组成必须与化学式相符，这是定量计算的基本依据。例如，测定 PO_4^{3-} 时，可以形成磷钼酸铵沉淀，但组成不固定，无法利用它作为测定 PO_4^{3-} 的称量形态。若采用磷钼酸喹啉法测定 PO_4^{3-}，则可得到组成与化学式相符的称量形态。

② 称量形要有足够的稳定性，不易吸收空气中的 CO_2、H_2O。例如测定 Ca^{2+} 时，若将 Ca^{2+} 沉淀为 $CaC_2O_4 \cdot H_2O$，灼烧后得到 CaO，易吸收空气中 H_2O 和 CO_2，因此，CaO 不宜作为称量形式。

③ 称量形态的摩尔质量要尽可能大，这样可增大称量形态的质量，以减

form and minimize weighing error. For instance, two weighing forms Al_2O_3 and $Al(C_9H_6NO)_3$ are used in the determination of aluminum ion respectively. If the mass of analyte Al is 0.1000g, 0.1888g of Al_2O_3 and 1.7040g of $Al(C_9H_6NO)_3$ can be produced respectively. The relative errors of two weighing forms are ±1% and ±0.1%. Obviously, weighing form $Al(C_9H_6NO)_3$ has higher accuracy than weighing form Al_2O_3 for the determination of Al.

小称量误差。例如在铝的测定中，分别用 Al_2O_3 和 8-羟基喹啉铝 [$Al(C_9H_6NO)_3$] 两种称量形态进行测定，若被测组分 Al 的质量为 0.1000g，则可分别得到 0.1888g Al_2O_3 和 1.7040g $Al(C_9H_6NO)_3$。两种称量形态由称量误差所引起的相对误差分别为 ±1% 和 ±0.1%。显然，以 $Al(C_9H_6NO)_3$ 作为称量形态比用 Al_2O_3 作态为称量形态测定铝含量的准确度高。

2. Factors Affecting Solubility of Precipitate

There are many factors affecting solubility, such as common ion effect, salt effect, acidic effect and complexing effect. In addition, temperature, solvent, precipitate structure and particle size can also influence solubility.

(1) Common ion effect The ion that forms precipitate crystal is called crystallization ion. The solubility of an ionic precipitate is reduced when an excess reagent or solution combining the crystallization ion is added to the solution in equilibrium with the precipitate. Such phenomenon is called **common ion effect**.

2. 影响沉淀溶解度的因素

影响沉淀溶解度的因素很多，如同离子效应、盐效应、酸效应、配位效应等。此外，温度、介质、沉淀结构和颗粒大小等对沉淀的溶解度也有影响。现分别进行讨论。

（1）同离子效应 组成沉淀晶体的离子称为构晶离子。当沉淀反应达到平衡后，如果向溶液中加入适当过量的含有某一构晶离子的试剂或溶液，则沉淀的溶解度减小，这种现象称为同离子效应。

For example:

At 25 ℃, the solubility of $BaSO_4$ in pure water is

$$s=[Ba^{2+}]=[SO_4^{2-}]= \sqrt{K_{sp}} = \sqrt{1.07\times10^{-10}} =1.03\times10^{-5}(mol \cdot L^{-1})$$

If SO_4^{2-} in solution increases to 0.10 mol · L^{-1}, then the solubility of $BaSO_4$ is

$$s=[Ba^{2+}]=K_{sp}/[SO_4^{2-}] =(1.07\times10^{-10}/0.10)\ mol \cdot L^{-1}=1.07\times10^{-9} mol \cdot L^{-1}$$

That is, the solubility of $BaSO_4$ decreases to ten thousandths.

Therefore in practical analysis, precipitating reagent is usually added in excess to create common ion effect to facilitate the complete precipitation of analyte. However, too much precipitating reagent migh cause side effects such as salt effect, acidic effect and complexing effect to increase solubility. In general, 50%-100% in excess is appropriate for volatile precipitating reagent, while 20-30% in excess is appropriate for non-volatile precipitating reagent.

因此，在实际分析中，常加入过量沉淀剂，利用同离子效应，使被测组分沉淀完全。但沉淀剂过量太多，可能引起盐效应、酸效应及配位效应等副反应，反而使沉淀的溶解度增大。一般情况下，沉淀剂过量 50%～100% 是合适的，如果沉淀剂是不易挥发的，则以过量 20%～30% 为宜。

(2) Salt effect　When precipitation reaction is in equilibrium, the solubility of precipitate increases if a strong electrolyte exists or other strong electrolytes are added. Such phenomenon is called salt effect. For instance, the solubility of AgCl and $BaSO_4$ in KNO_3 solution is greater than that in pure water. The greater the KNO_3 concentration is, the higher the solubility.

(3) Acidic effect　The influence of acidity of solution on solubility of precipitate is called acidic effect. The occurrence of acidic effect is mainly due to the influence of the degree of H^+ concentration on the dissociation equilibrium of weak acid, polybasic acid or sparingly soluble acid. Therefore, acidic effect has different influences on different types of precipitate. If the precipitate is strong acid salt (e. g. $BaSO_4$, AgCl), acidity has slight influence on its solubility. However, if the precipitate is weak acid salt (e. g. CaC_2O_4), acidity has obvious influence on its solubility.

(4) Coordination effect　In precipitation reaction, if there exists in solution the complexing agent that can produce soluble complex with crystallization ion, the solubility of precipitate will be increased. Such phenomenon is called coordination effect. Complexing agent comes from two sources: precipitate itself is a complexing agent; the other reagents that added in the solution.

In a word, practical conditions have to be considered to know which effect prevails in the reaction. For strong acid precipitate without complexing reaction, common ion effect and salt effect may occur. For weak acid precipitate or sparingly soluble salt, acidic effect occurs in most cases. For precipitate having complexing reaction and with high solubility product, stable complex is easily produced and complexing effect may occur. Aside from the factors stated above, temperature, other solvents, particle size and structure of precipitate may also affect solubility of precipitate.

3. Procedure for Precipitation Gravimetry

The basic operation procedure of precipitation gravimetry

includes dissolution, precipitation, filtering, washing, drying and ignition. They are introduced as follows.

(1) Dissolution Sample is weighed and put into a beaker. Solvent is added into the beaker along the wall. Cover the beaker with a watch glass and shake slightly. If necessary, heat to increase solubility. But the temperature should not be too high to prevent the solution from splashing and losing.

(2) Precipitation The requirement for precipitate in gravimetry analysis is complete precipitation and pure composition. In order to meet this requirement, proper precipitating conditions must be chosen in accordance with different types of precipitates, such as volume and temperature of solution, concentration, quantity, adding rate, stirring speed and settling time of precipitating reagent. The specific procedure has to be followed.

In the operation of precipitation, operate the buret with left hand to add precipitating reagent dropwisely. In the mean time, stir the solution with a glass rod in right hand. The glass rod should not touch the wall or bottom of beaker to prevent damage during stirring. The solution is then heated in a water bath or on a hot plate. Afterwards, check if the precipitation is complete in the following way: when the precipitate settles down, add one drop of precipitating reagent along the wall of the beaker into the supernatant clarified liquid to see if any turbidity occurs. If tuibidity does not occur, it shows that the precipitation is complete. If not, add more precipitating reagent. Then repeat the above checking procedure utill no turbidity occurs. Cover the beaker with a watch glass.

(3) Filtering and washing

① Filtering with filter paper

a. Choosing of filter paper Filter paper is classified into two types: quanlitative filter paper and quantitative

解、沉淀、过滤、洗涤、干燥和灼烧等步骤，分别介绍如下。

（1）溶解样品 称量样品置于烧杯中，沿杯壁加溶剂，盖上表面皿，轻轻摇动，必要时可加热使其溶解，但温度不可太高，以防溶液溅失。

（2）沉淀 重量分析对沉淀的要求是尽可能地完全和纯净，为了达到这个要求，应该按照沉淀的不同类型选择不同的沉淀条件，如沉淀时溶液的体积、温度，加入沉淀剂的浓度、数量，加入速度，搅拌速度，放置时间等。因此，必须按照规定的操作程序进行。

一般进行沉淀操作时，左手拿滴管，滴加沉淀剂，右手持玻璃棒不断搅动溶液，搅动时玻璃棒不要碰到烧杯壁或烧杯底，以免划损烧杯。溶液需要加热时，一般在水浴或电热板上进行，沉淀后应检查沉淀是否完全，检查的方法是：待沉淀下沉后，在上层澄清液中，沿烧杯壁加1滴沉淀剂，观察滴落处是否出现浑浊，无浑浊出现表明已沉淀完全，如出现浑浊，需再补加沉淀剂，直至再次检查时上层清液中不再出现浑浊为止，然后盖上表面皿。

（3）过滤和洗涤

① 用滤纸过滤

a. 滤纸的选择 滤纸分定性滤纸和定量滤纸两种，重量分析中常用定

Procedure for precipitation gravimetry: dissolution and precipitation
沉淀重量法的基本操作之样品溶解与沉淀制备

Procedure for precipitation gravimetry: filtering with filter paper
沉淀重量法的基本操作之沉淀过滤（滤纸过滤）

filter paper. The latter type (also called ashless filter paper) is generally used in the gravimetry analysis, which leaves very little amount of ash after ignition so that its weight can be neglected. If ash is heavy, deduct the blank value. Generally, quantitative filter paper is circular, with diameters of 11cm, 9cm and 7cm. It can also be classified into rapid, medium and slow flow rate papers according to the size of pore. Choose proper filter paper on the basis of the property of precipitate. For example, fine crystalline precipitates (e. g. $BaSO_4$, $CaC_2O_4 \cdot 2H_2O$) choose slow flow rate paper; amorphous precipitates (e. g. $Fe_2O_3 \cdot nH_2O$) choose rapid flow rate paper; coarse crystalline precipitates (e. g. $MgNH_4PO_4$) choose medium flow rate paper. In the meanwhile, choose proper size of filter paper on the basis of the quantity of precipitate.

b. Choosing of funnel Long stem funnel is used in gravimetric analysis, with stem 15-20cm long, cone vertex angle 60°. The diameter of stem should be small, 3-5mm in usual so that water column can be kept in the stem. The outlet of funnel is ground into 45°, see figure 6-1. The funnel must be thoroughly cleaned before use.

c. Folding of filter paper Clean hands and wipe off water before folding filter paper. The folding of filter paper is shown in figure 6-2.

Double over the filter paper and press down half of it. Then fold the half in half again but do not press. Place the folded conical filter paper into funnel. The filter

量滤纸（或称无灰滤纸）进行过滤。定量滤纸灼烧后灰分极少，其重量可忽略不计，如果灰分较重，应扣除空白。定量滤纸一般为圆形，按直径分有11cm、9cm、7cm等几种；按滤纸孔隙大小分有"快速""中速"和"慢速"3种。根据沉淀的性质选择合适的滤纸，如$BaSO_4$、$CaC_2O_4 \cdot 2H_2O$等细晶形沉淀，应选用"慢速"滤纸过滤；$Fe_2O_3 \cdot nH_2O$为胶状沉淀，应选用"快速"滤纸过滤；$MgNH_4PO_4$等粗晶形沉淀，应选用"中速"滤纸过滤。根据沉淀量的多少，选择滤纸的大小。

b. 漏斗的选择 用于重量分析的漏斗应该是长颈漏斗，颈长为15～20cm，漏斗锥体角应为60°，颈的直径要小些，一般为3～5mm，以便在颈内容易保留水柱，出口处磨成45°角，如图6-1所示。漏斗在使用前应洗净。

c. 滤纸的折叠 折叠滤纸的手要洗净擦干。滤纸的折叠如图6-2所示。

先把滤纸对折并按紧一半，然后再对折但不要按紧，把折成圆锥形的滤纸放入漏斗中。滤纸的大小应低于漏斗边

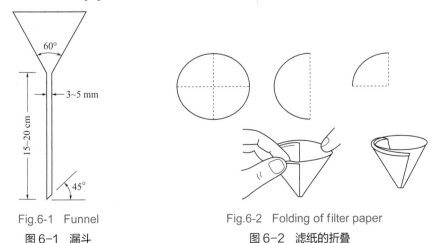

Fig.6-1 Funnel
图6-1 漏斗

Fig.6-2 Folding of filter paper
图6-2 滤纸的折叠

paper should be 0.5-1cm lower than the edge of funnel. If not, cut a little. Check if the folded filter paper fits tighly with the inner wall of funnel. If it's not, adjust the folding angle until it completely fits with funnel, then press down the part folded in the second time. Take out the filter paper cone, tear off a piece from the half with three layers so that the inner layers can stick to the inner wall of funnel. The torn-off piece will be used to wipe off the residual precipitate in the beaker.

d. Making water column　　Take a washing bottle and wet the whole filter paper. Drive all the air bubbles between filter paper and funnel wall by slightly pressing filter paper with fingers. Feed in water till the edge of paper so that the funnel stem is fully filled with water to form a water column. The water column shall still remains when water on the filter water drains off completely. In this way, the gravity of water plays a function of suction to speed up filtering.

If the making of water column fails in the above operation, plug the outlet of funnel with a finger and slightly lift up a side of the filter paper. Fill water into the gap between filter paper and funnel by a washing bottle till the funnel stem and part of funnel cone are full of water. Press back the filter paper down to the wall of funnel while with drawing the finger that is plugging the funnel outlet. At this time, the water column should have been made. However, if water column still cannot form or it cannot keep, in the mean time, the funnel stem is absolutely clean, the funnel has to be replaced by another one with narrower stem. It has been found that the funnel with large stem cannot make water column.

Place the funnel that has been made water column on a funnel support above a clean beaker for collecting filtrate. The filtrate can be used to determine other compositions. Even if the filtrate is not needed, a clean beaker still has to be placed beneath the funnel in consideration of the percolation of precipitate or accidental damage of filter paper so that filtration needs to be redone. To prevent splattering of filtrate, the longer side of funnel outlet usually abuts against the inner wall of beaker. The

缘 0.5～1cm 左右，若高出漏斗边缘，可剪去一圈。观察折好的滤纸是否能与漏斗内壁紧密贴合，若未贴合紧密可以适当改变滤纸折叠角度，直至与漏斗贴紧后把第二次的折边折紧。取出圆锥形滤纸，将半边为三层滤纸的外层折角撕下一块，这样可以使内层滤纸紧密贴在漏斗内壁上，撕下来的那一小块滤纸保留作擦拭烧杯内残留的沉淀用。

d. 做水柱　　滤纸放入漏斗后，用手按紧使之密合，然后用洗瓶加水润湿全部滤纸。用手指轻压滤纸赶去滤纸与漏斗壁间的气泡，然后加水至滤纸边缘，此时漏斗颈内应全部充满水，形成水柱。当滤纸上的水全部流尽后，漏斗颈内的水柱应仍能保住，这样，由于液体的重力可起抽滤作用，加快过滤速度。

若水柱未做成，可用手指堵住漏斗下口，稍掀起滤纸的一边，用洗瓶向滤纸和漏斗间的空隙内加水，直到漏斗颈及锥体的一部分被水充满，然后边按紧滤纸边慢慢松开下面堵住出口的手指，此时水柱应该形成。如仍不能形成水柱，或水柱不能保持，而漏斗颈又确已洗净，则是因为漏斗颈太大。实践证明，漏斗颈太大的漏斗，是做不出水柱的，应更换漏斗。

做好水柱的漏斗应放在漏斗架上，下面用一个洁净的烧杯承接滤液，滤液可用作其他组分的测定。滤液有时是不需要的，但考虑到过滤过程中，可能有沉淀渗滤，或滤纸意外破裂，需要重滤，所以要用洗净的烧杯来承接滤液。为了防止滤液外溅，一般都将漏斗颈出口斜口长的一侧贴紧烧杯内壁。漏斗位置的高低，以过滤过程中漏斗颈的出口不接触滤液为度。

e. 倾泻法过滤和初步洗涤　　首先要强调，过滤和洗涤一定要一次完成，因

position of funnel depends on the rule that the funnel outlet does not touch filtrate during filtration.

e. Decantation filtration and preliminary washing It has to be emphasized that filtering and washing must be done in one time. Therefore, time should be estimated in advance to avoid interruption, especially for the filtration of amorphous precipitate.

Filtration is generalized into three stages: first stage, filter as much as possible the supernatant by decantation filtration and premilinarily wash the precipitate in the beaker; second stage, move the precipitate onto funnel; third stage, clean beaker and wash the precipitate on funnel.

Decantation filtration is often applied to avoid plugging of filter paper pores by precipitate and affect filtration speed. That is, leave the beaker to stand slantingly, waiting for the precipitate settling down. Pour the supernatant into funnel first in lieu of the mixture of precipitate and solution at the beginning of filtration.

The operation of filtration is shown in figure 6-3. Move the beaker above the funnel and lift up the glass rod carefully. Let the lower end of glass rod touch the wall of beaker so that the drops pending on the rod could run back into beaker. Abut the glass rod against the beaker

Procedure for precipitation gravimetry:
decantation filtration
沉淀重量法的基本操作之倾泻法过滤

此必须事先计划好时间，不能间断，特别是过滤胶状沉淀。

过滤一般分3个阶段进行：第一阶段采用倾泻法把尽可能多的清液先过滤出去，并对烧杯中的沉淀作初步洗涤；第二阶段把沉淀转移到漏斗上；第三阶段清洗烧杯和洗涤漏斗上的沉淀。

过滤时，为了避免沉淀堵塞滤纸的空隙，影响过滤速度，一般多采用倾泻法过滤，即倾斜静置烧杯，待沉淀下降后，先将上层清液倾入漏斗中，而不是一开始过滤就将沉淀和溶液搅混后过滤。

过滤操作如图6-3所示，将烧杯移到漏斗上方，轻轻提取玻璃棒，将玻璃棒下端轻碰一下烧杯壁使悬挂的液滴流回烧杯中，将烧杯嘴与玻璃棒贴紧，玻璃棒直立，下端接近三层滤纸的一边，

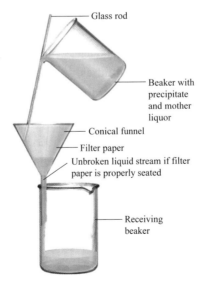

Fig. 6-3 Filtering a precipitate (The conical funnel is supported by a metal ring attached to a ring stand, neither of which is shown.)

图6-3 过滤沉淀（锥形漏斗由连接在环形支架上的金属环支撑，图中两者均未显示。）

spout. The glass rod is held vertically with lower end touching the three-layer side of filter paper. Incline the beaker slowly to let the supernatant flow into funnel along the glass rod. The level in the funnel should not be higher than two third of the height of filter paper, or be 5mm lower than the upper edge of filter paper to prevent the loss due to the capillary action in which precipitate flows over the upper edge of filter paper.

When decanting pauses, lift the beaker spout up along the glass rod till the beaker gradually stays vertically. When the mutual position of the glass rod and beaker changes from perpendicular to almost parallel, take the glass rod away from the spout and put it into the beaker so that the liquid pending on the end of glass and on the spout does not flow onto the outer wall of beaker. When putting the glass rod back into the beaker, neither stir up the supernatant nor lean the glass rod against the beaker spout because of a small amount of precipitate left on the spout. Repeat above operation utill all the supernatant is decanted. When there is only less liquid lcft in the beaker, making it difficult to pour, incline the glass rod toward left to increase the tilting angle of beaker.

During washing, inject a small amount of washing liquid (about 20 mL each time) around the inner wall of beaker. Stir thoroughly and then let the mud settle until getting precipitate. Decant and filter in accordance with the above procedures. Repeat washing and precipitating four or five times. Pour out the washing liquid as completely as possible each time, then inject in fresh washing liquid for the next time. Always check if filtrate is clear free from any precipitate particles. If not, redo the filtering or redo the experiment.

f. Transferring of precipitate　　After precipitate is washed via decantation, add a little washing liquid into the beaker with precipitate. Stir and mix, then pour all the mixture into funnel. Repeat this procedure two or three times. Put the glass rod over the rim of beaker horizontally. The lower end of the rod protrudes 2-3cm from the spout of beaker. The forefinger of the left hand presses on the glass rod, the thumb is in the front of

慢慢倾斜烧杯，使上层清液沿玻璃棒流入漏斗中，漏斗中的液面不要超过滤纸高度的 2/3。或使液面离滤纸上边缘约 5mm，以免少量沉淀因毛细管作用越过滤纸上缘，造成损失。

暂停倾注时，应沿玻璃棒将烧杯嘴往上提，逐渐使烧杯直立，等玻璃棒和烧杯由相互垂直变为几乎平行时，使玻璃棒离开烧杯嘴而移入烧杯中。这样才能避免留在棒端及烧杯嘴上的液体流到烧杯外壁上去。玻璃棒放回原烧杯时，勿将清液搅混，也不要靠在烧杯嘴处，因嘴处沾有少量沉淀。如此重复操作，直至上层清液倾完为止。当烧杯内的液体较少而不便倾出时，可将玻璃棒稍向左倾斜，使烧杯倾斜角度更大些。

洗涤时，沿烧杯内壁四周注入少量洗涤液，每次约 20mL 左右，充分搅拌，静置，待沉淀沉降后，按上法倾注过滤，如此洗涤沉淀 4～5 次，每次应尽可能把洗涤液倾倒尽，再加第二份洗涤液。随时检查滤液是否透明不含沉淀颗粒，否则应重新过滤，或重做实验。

f. 沉淀的转移　　沉淀用倾泻法洗涤后，在盛有沉淀的烧杯中加入少量洗涤液，搅拌混合，全部倾入漏斗中。如此重复 2～3 次，然后将玻璃棒横放在烧杯口上，玻璃棒下端比烧杯口长出 2～3cm，左手食指按住玻璃棒，大拇

Procedure for precipitation gravimetry: transferiing of precipitate
沉淀重量法的基本操作之沉淀转移

beaker while the rest fingers are at the back of beaker. Take up the beaker and hold it over the funnel. Tilt the beaker and let the glass rod still point to the side of filter paper with three layers. Flush the inner wall of beaker with washing bottle to transfer the precipitate stuck to the wall into funnel, as shown in figure 6-4. Finally, wipe the glass rod with the piece of filter paper left before. Then put that piece of filter paper into beaker to wipe the wall of beaker by pressing it down with the glass rod. Slide the piece of paper into funnel with the glass rod after wiping. Flush the beaker again with washing liquid to transfer all the precipitate residue into funnel.

g. Washing　　When all the precipitate is transferred onto the filter paper, do the final washing on the filter paper. Move the washing bottle spirally down from the the place a little lower than the edge of filter paper to flush precipitate, as shown in figure 6-5. In this way, precipitate can be flushed down to the bottom of filter paper cone. Do not flush the washing liquid into the precipitate in the center of filter paper to avoid splattering of precipitate.

The washing of precipitate adopts the method of "frequent, small amount". That is, flush with a small amount of washing liquid every time and drain well after washing. Then flush with washing liquid for the second time so as to improve washing efficiency. The frequency of washing is also stipulated. It is generally required to wash eight to ten times.

指在前，其余手指在后，拿起烧杯，放在漏斗上方，倾斜烧杯使玻璃棒仍指向三层滤纸的一边，用洗瓶冲洗烧杯壁上附着的沉淀，使之全部转移入漏斗中，如图 6-4 所示。最后用保存的小块滤纸擦拭玻璃棒，再放入烧杯中，用玻璃棒压住滤纸进行擦拭。擦拭后的滤纸块，用玻璃棒拨入漏斗中，用洗涤液再次冲洗烧杯，将残存的沉淀全部转入漏斗中。

g. 洗涤　　沉淀全部转移到滤纸上后，再在滤纸上进行最后的洗涤。这时要用洗瓶由滤纸边缘稍下一些地方螺旋形向下移动冲洗沉淀，如图 6-5 所示。这样可使沉淀集中到滤纸锥体的底部，不可将洗涤液直接冲到滤纸中央沉淀上，以免沉淀外溅。

Procedure for precipitation gravimetry: washing
沉淀重量法的基本操作之洗涤与检验

采用"少量多次"的方法洗涤沉淀，即每次加少量洗涤液，洗后尽量沥干，再加第二次洗涤液，这样可提高洗涤效率。洗涤次数也有规定，一般洗涤 8~10 次。

Fig.6-4　Washing off the final amount of precipitate
图 6-4　洗脱最后的沉淀

Fig.6-5　Washing and precipitating
图 6-5　洗涤沉淀

② Filtering with micropore glass crucible (funnel)　Some precipitates cannot be filtered via filter paper because of the great change in weight when the filter paper dries. In these cases, micropore glass crucible (or micropore glass funnel) is used (fig.6-6). The filter plate of micropore glass crucible (funnel) is manufactured by sintering glass powder under high temperature.

(4) Drying and ignition　The drying and ignition of precipitate will be carried out in a crucible pre-dried to a constant weight. Therefore, preparation of crucible is necessary before the drying and ignition of precipitate.

① Preparation of crucible　Clean the porcelain crucible, dry with soft fire and number it (number on the outer wall of crucible with blue ink containing Fe^{3+} or Co^{2+}). Heat crucible to the required temperature. The heating is done in a high temperature electric oven. Because sudden rise or drop in temperature often causes breakage of crucible, it is better to place the crucible into oven when it is still at ambient temperature, or preheat the crubile at the hole of oven which has already been at a high temperature. The crucible is usually fired at 800-950℃ for half an hour (one hour for new crucibles). Before taking out the crucible from oven, cool down the oven first, then move the crucible into a desiccator.

② 用微孔玻璃坩埚（漏斗）过滤　有些沉淀不能用滤纸过滤，因为滤纸烘干后，重量改变很多，在这种情况下，应该用微孔玻璃坩埚（或微孔玻璃漏斗）过滤，如图 6-6 所示。这种过滤器的滤板是用玻璃粉末在高温熔结而成的。

（4）干燥和灼烧　沉淀的干燥和灼烧是在一个预先灼烧至质量恒定的坩埚中进行的。因此，在沉淀的干燥和灼烧前，必须预先准备好坩埚。

① 坩埚的准备　先将瓷坩埚洗净，小火烤干或烘干，编号（可用含 Fe^{3+} 或 Co^{2+} 的蓝墨水在坩埚外壁上编号），然后在所需温度下，加热灼烧。灼烧可在高温电炉中进行。由于温度骤升或骤降常使坩埚破裂，最好将坩埚放入冷的炉膛中逐渐升高温度，或者将坩埚在已升至较高温度的炉膛口预热一下，再放进炉膛中。一般在 800～950℃下灼烧半小时（新坩埚需灼烧 1h）。从高温炉中取出坩埚时，应先使高温炉降温，然后将坩埚移入干燥器中，将干燥器连同坩埚一起移至天平室，冷却至室

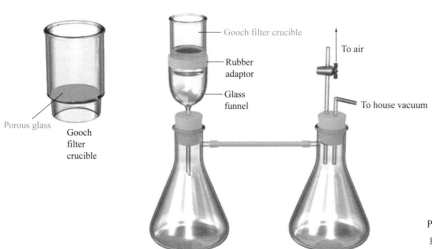

Fig. 6-6　Micropore glass crucible and micropore glass funnel

图 6-6　微孔玻璃坩埚和微孔玻璃漏斗

Procedure for precipitation gravimetry: filtering with micropore glass crucible
沉淀重量法的基本操作之微孔玻璃坩埚的使用

Move the desiccator together with the crucible into balance room to let it cool for about 30 minutes to room temperature. Take the crucible out and weigh. Then the crucible is to be fired for the second time for 15-20 minutes, and then cooled and weighed. When the mass difference between two weighings is not more than 0.2mg, the crucible reaches constant mass. Otherwise, the firing, cooling and weighing have to be repeated. The preheating temperature of crucible should be same with the ignition temperature of precipitate later.

② Drying and ignition of precipitate　Start the drying and ignition of precipitate after the crucible is ready. Take out filter paper and precipitate from funnel with the glass rod. Fold and roll them into a packet in which the precipitate is contained, as shown in figure 6-7 and figure 6-8. Special attention should be paid that precipitate does not have any loss during the operation. If there is a little precipitate stuck to the funnel, wipe it off with a piece of filter paper and fold it into the packet too.

Put the packet into the crucible. The side with multi-layer filter paper faces upwards so that the ashing of filter paper can be easily done. Place the crucible with lid on a clay triangle in a inclined position, as shown in figure 6-9. Dry and carbonize the filter paper as shown in figure 6-10. Care must be taken to prevent burning

温（约需 30min），取出称量。随后进行第二次灼烧，约 15～20min，冷却和称量。如果前后两次称量结果之差不大于 0.2mg，即可认为坩埚已达质量恒定，否则还需再次灼烧，直至质量恒定为止。灼烧空坩埚的温度必须与以后灼烧沉淀的温度一致。

② 沉淀的干燥和灼烧　坩埚准备好后，即可开始沉淀的干燥和灼烧。利用玻璃棒把滤纸和沉淀从漏斗中取出，如图 6-7、图 6-8 所示，折卷成小包，把沉淀包卷在里面。此时应特别注意，勿使沉淀有任何损失。如果漏斗上沾有些微沉淀，可用滤纸碎片擦下，与沉淀包卷在一起。

Procedure for precipitation gravimetry: drying and ignition of precipitate
沉淀重量法的基本操作之沉淀的干燥和灼烧

将滤纸包装进已质量恒定的坩埚内，使滤纸层较多的一边向上，可使滤纸较易灰化。如图 6-9 所示，倾斜坩埚于泥三角上，盖上坩埚盖，然后如图 6-10 所示，将滤纸烘干并炭化。在此

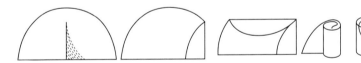

Fig.6-7　Folding of filter paper after filtration
图 6-7　过滤后的滤纸折叠

Fig.6-8　Folding of filter paper with amorphous precipitate
图 6-8　无定形沉淀的滤纸折叠

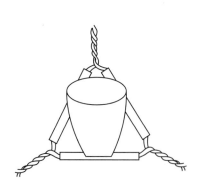

Fig.6-9 Crucible inclinely placed on a clay triangle

图6-9 坩埚倾斜放置在泥三角上

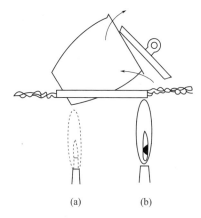

Fig.6-10 Drying (b) and carbonization (a)

图6-10 烘干（b）炭化（a）

of filter paper in case precipitate flies off causing mass loss. If the filter paper is on fire, move away the Bensen burner immediately. At the same time, cover crucible with lid to let the fire extinguish by itself.

After the filter paper is carbonized, increase the temperature gradually while turning crucible with tongs. All the black carbon on the inner wall of crucible should be completely ignited. The procedure in which carbon is ignited into CO_2 and removed is called **ashing**. When the filter paper is fully ignited, place the crucible on the clay triangle vertically and cover with lid (leaving a slit). Dry the precipitate under certain temperature or in the high temperature electric oven. In general, it takes about 30-45 minutes for the first firing and 15-20 minutes for the second. The crucible should be taken out of the oven and cool in the air after each firing before being moved into a desiccator. After the precipitate is cooled to room temperature, weigh it. Repeat the firing, cooling and weighing until the crucible reaches constant mass.

The micropore glass crucible (funnel) needs only drying before weighing. Place precipitate together with micropore glass crucible on a watch glass and dry in an air oven. The drying temperature is set according to the property of the precipitate. In principle, the first drying takes longer time, about 2 hours; but the second drying takes shorter time, about 45 minutes to 1h. The actual

过程中必须防止滤纸着火，否则会使沉淀飞散而损失。若已着火，应立刻移开煤气灯，并将坩埚盖盖上，让火焰自熄。

当滤纸炭化后，可逐渐提高温度，并随时用坩埚钳转动坩埚，把坩埚内壁上的黑炭完全烧去。将炭烧成 CO_2 而除去的过程叫灰化。待滤纸灰化后，将坩埚垂直地放在泥三角上，盖上坩埚盖（留一小孔隙），于指定温度下灼烧沉淀，或者将坩埚放在高温电炉中灼烧。一般第一次灼烧时间为 30～45min，第二次灼烧 15～20min。每次灼烧完毕从炉内取出后，都需要在空气中稍冷，再移入干燥器中。沉淀冷却到室温后称量，然后再灼烧、冷却、称量，直至质量恒定。

微孔玻璃坩埚（或漏斗）只需烘干即可称量，一般将微孔玻璃坩埚（或漏斗）连同沉淀放在表面皿上，然后放入烘箱中，根据沉淀性质确定烘干温度。一般第一次烘干时间要长些，约 2h，第二次烘干时间可短些，约 45min 到 1h，根据沉淀的性质具体处理。沉淀烘

drying time also depends on the perperty of precipitate. When the drying is complete, take the crucible (funnel) out and put it into a desiccator to cool to room temperature. Then weigh it. Repeat the drying, cooling and weighing until the crucible reaches constant mass.

③ Usage of desiccator　A desiccator is a sealable heavy glass container with a ground-glass rim of lid used for preserving items such as crucibles, weighing bottles and samples. The ground-glass rim of the lid is greased with a thin layer of petroleum jelly to seal the lid with the container, see figure 6-11.

The lower compartment of the desiccator contains desiccant. The most commonly used desiccants are allochroic silica gel and anhydrous calcium chloride. A clean perforated porcelain platform is placed over desiccants. Crucibles can be placed in the pores of the platform.

The water absorption ability of desiccants is limited. For example, at 20℃, the residual vapor in 1L of air which has been dried by silica gel is 6×10^{-3}mg; at 25℃, the residual vapor in 1L of air which has been dried by anhydrous calcium chloride is less than 0.36mg. It means the air in the desiccator is not absolutely dry. There is humidity but very low.

干后，取出坩埚（或漏斗），置于干燥器中冷却至室温后称量。反复烘干、称量，直至质量恒定为止。

③ 干燥器的使用方法　干燥器是具有磨口盖子的密闭厚壁玻璃器皿，常用于保存坩埚、称量瓶、试样等物。它的磨口边缘涂一薄层凡士林，使之能与盖子密合，如图6-11所示。

干燥器底部盛放干燥剂，最常用的干燥剂是变色硅胶和无水氯化钙，其上搁置洁净的带孔瓷板。坩埚等即可放在瓷板孔内。

干燥剂吸收水分的能力是有限的。例如硅胶，20℃时，被其干燥过的1L空气中残留水分为6×10^{-3}mg；无水氯化钙，25℃时，被其干燥过的1L空气中残留水分小于0.36mg。因此，干燥器中的空气并不是绝对干燥的，只是湿度较低而已。

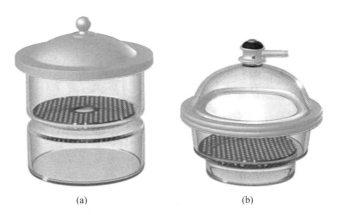

(a)　　　　　　　(b)

Fig. 6-11　(a) Ordinary desiccator. (b) Vacuum desiccator that can be evacuated through the side arm at the top and then sealed by rotating the joint containing the side arm. Drying is more efficient at low pressure.

图6-11　（a）普通干燥器。（b）真空干燥器，可以通过顶部的侧臂抽真空，然后通过旋转包含侧臂的接头进行密封，其在低压下干燥更有效。

The following cautions should be made when using desiccators:

a. Desiccants should not be much in case the lower compartment of desiccator is contaminated.

b. Use both hands to move the desiccator with thumbs holding down the lid, as shown in figure 6-12.

c. Do not open the lid upwards. Remove the lid by pressing the desiccator with left hand and push the lid sideways horizontally. Let cold air flow into the desiccator, then open the desiccator completely. The lid must be placed inside upwards on the table.

d. Do not place hot substances into desiccators.

e. If a hotter substance is put into desiccator, the expanded air in the desiccator may lift the lid up. To prevent the damage of lid, press the lid down with a hand. Slide the lid sideways from time to time (for one second) to let out the hot air.

f. The dried and ignited crucible and precipitate cannot be placed in the desiccator for a long time. Otherwise, they will absorb water increasing weight.

g. Allochroic silica gel is blue (containing anhydrous Co^{2+}) and turns pink (containing hydrous Co^{2+}) when it bonds with water. It can be turned into blue silica gel when dried at 120℃ for reuse until it is broken and not suitable for use.

使用干燥器时应注意下列事项：

a. 干燥剂不可放得太多，以免沾污坩埚底部。

b. 搬移干燥器时，要用双手拿着，用大拇指紧紧按住盖子，如图6-12所示。

c. 打开干燥器时，不能往上掀盖，应用左手按住干燥器，右手小心地把盖子稍微推开，等冷空气徐徐进入后，才能完全推开，盖子必须仰放在桌面上。

d. 不可将太热的物体放入干燥器中。

e. 有时较热的物体放入干燥器中后，空气受热膨胀会把盖子顶起来，为了防止盖子被打翻，应当用手按住，不时把盖子稍微推开（不到1s），以放出热空气。

f. 灼烧或烘干后的坩埚和沉淀，在干燥器内不宜放置过久，否则会因吸收一些水分而使质量略有增加。

g. 变色硅胶干燥时为蓝色（无水 Co^{2+} 色），受潮后变粉红色（水合 Co^{2+} 色）。可以在120℃烘干受潮的硅胶待其变蓝后反复使用，直至破碎不能用为止。

Fig.6-12 Moving desiccator

图6-12 干燥器的搬移

Section 4 Calculation of Analytical Result

第4节 重量分析结果计算

1. Conversion Factor in Gravimetric Analysis

1. 重量分析中的换算因数

In gravimetric analysis, it is easy to calculate the analytical result when the formula of the final weighing form is the same as that of analyte. For example, it is required to calculate the content of SiO_2. When the final weighing form is also SiO_2, the analytical result can be calculated as follows:

重量分析中，当最后称量形态与待测组分形式一致时，计算其分析结果就比较简单了。例如，测定要求计算 SiO_2 的含量，重量分析最后称量形态也是 SiO_2，其分析结果见式（6.3）。

$$w_{SiO_2} = \frac{m_{SiO_2}}{m_s} \times 100\% \tag{6.3}$$

In which, w_{SiO_2} is the mass fraction of SiO_2 (value shown as %); m_{SiO_2} is the mass of SiO_2 precipitate, g; m_s is the mass of sample, g.

式中，w_{SiO_2} 为 SiO_2 的质量分数（数值以%表示）；m_{SiO_2} 为 SiO_2 沉淀质量，g；m_s 为试样质量，g。

However, if the formula of the final weighing form is different from that of analyte, the analytical result should be converted. For instance, 0.5051g of $BaSO_4$ precipitate is got when determining Ba, convert the mass of analyte Ba in the following way:

如果最后称量形态与待测组分形式不一致时，分析结果就要进行适当的换算。如测定钡含量时，得到 $BaSO_4$ 沉淀0.5051g，可按下列方法换算成被测组分钡的质量。

$$BaSO_4 \longrightarrow Ba$$
$$233.4 \qquad\qquad 137.4$$
$$0.5051g \qquad\qquad m_{Ba}$$
$$m_{Ba} = 0.5051 \times 137.4/233.4 \text{g} = 0.2973\text{g}$$

that is,

$$m_{Ba} = m_{BaSO_4} \frac{M(Ba)}{M(BaSO_4)} \tag{6.4}$$

In which, m_{BaSO_4} is the mass of weighing form $BaSO_4$, g; $\frac{M(Ba)}{M(BaSO_4)}$ is the fraction converting the mass of $BaSO_4$ into the mass of Ba, which is constant, independent of the mass of sample. This ratio is generally called **gravimetric factor** or **chemical factor** (that is, the ratio

式中，m_{BaSO_4} 为称量形态 $BaSO_4$ 的质量，g；$\frac{M(Ba)}{M(BaSO_4)}$ 是将 $BaSO_4$ 的质量换算成 Ba 的质量的分式，此分式是一个常数，与试样质量无关，这一比值通常称为换算因数或化学因数（即待测

of the molar mass of analyte with the molar mass of weighing form, expressed with F). After converting the mass of weighing form into the mass of analyte, calculate as in the example of SiO_2.

When solving the gravimetric factor, note that the number of atom or molecule of analyte in numerator and denominator should be same. Therefore, sometime a coefficient is multiplied before the molar mass of analyte and weighing form respectively. The gravimetric factors of some common substances can be found out in the manual of analytical chemistry. In table 6-1, the gravimetric factors of a few common substances are listed.

组分的摩尔质量与称量形态的摩尔质量之比,常用 F 表示)。将称量形态的质量换算成所要测定组分的质量后,即可按前面计算 SiO_2 分析结果的方法进行计算。

求换算因数时,一定要注意使分子和分母所含被测组分的原子或分子数目相等,所以在待测组分的摩尔质量和称量形态摩尔质量之前有时需要乘以适当的系数。分析化学手册中可查到常见物质的换算因数。表 6-1 列出几种常见物质的换算因数。

Table 6-1 Gravimetric factor of a few common substances
表6-1 几种常见物质的换算因数

Analyte 待测物质	Precipitation Form 沉淀形态	Weighing Form 称量形态	Gravimatric Factor 换算因数
Fe	$Fe_2O_3 \cdot nH_2O$	Fe_2O_3	$2M(Fe)/M(Fe_2O_3)=0.6994$
Fe_3O_4	$Fe_2O_3 \cdot nH_2O$	Fe_2O_3	$2M(Fe_3O_4)/3M(Fe_2O_3)=0.9666$
P	$MgNH_4PO_4 \cdot 6H_2O$	$Mg_2P_2O_7$	$2M(P)/M(Mg_2P_2O_7)=0.2783$
P_2O_5	$MgNH_4PO_4 \cdot 6H_2O$	$Mg_2P_2O_7$	$M(P_2O_5)/M(Mg_2P_2O_7)=0.6377$
MgO	$MgNH_4PO_4 \cdot 6H_2O$	$Mg_2P_2O_7$	$2M(MgO)/M(Mg_2P_2O_7)=0.3621$
S	$BaSO_4$	$BaSO_4$	$M(S)/M(BaSO_4)=0.1374$

2. Examples of Calculation

2. 结果计算示例

Example 6-1

Determination of sulfur content in iron ores with bariun sulfate gravimetric method. The mass of sample is 0.1819g, and the mass of the final $BaSO_4$ precipitate is 0.4821g. What is the mass fraction of sulfur in the sample.

Solution

The precipitate is barium sulfate while the weighing form is also barium sulfate. However, the analyte is sulfur. Therefore weighing form must be converted into the analyte by a factor so as to calculate the content of analyte. The relative molecular mass of barium sulfate is 233.4; the relative atomic mass of sulfur is 32.06.

$$w_s = \frac{m_S}{m_s} \times 100\% = \frac{m_{BaSO_4} \frac{M(S)}{M(BaSO_4)}}{m_s} \times 100\%$$

So
$$w_s = \frac{0.4821 \times 32.06/233.4}{0.1819} \times 100\% = 36.41\%$$

Example 6-2

Take 0.1666g of sample to determine the ferrum content in magnetite (impure Fe_3O_4). The sample is dissolved and oxidized, leaving a Fe^{3+} precipitate $Fe(OH)_3$, which is ignited into 0.1370g Fe_2O_3. Calculate in the sample (1) mass fraction of Fe, (2) mass fraction of Fe_3O_4.

Solution

(1) $M(Fe)= 55.85$ g·mol^{-1}; $M(Fe_3O_4)=231.5$ g·mol^{-1}; $M(Fe_2O_3)=159.7$ g·mol^{-1}

$$w_{Fe} = \frac{m_{Fe}}{m_s} \times 100\% = \frac{m_{Fe_2O_3} \frac{2M(Fe)}{M(Fe_2O_3)}}{m_s} \times 100\%$$

So
$$w_{Fe} = \frac{0.1370 \times 2 \times 55.85/159.7}{0.1666} \times 100\% = 57.52\%$$

(2)
$$w_{Fe_3O_4} = \frac{m_{Fe_3O_4}}{m_s} \times 100\% = \frac{m_{Fe_2O_3} \frac{2M(Fe_3O_4)}{3M(Fe_2O_3)}}{m_s} \times 100\%$$

So
$$w_{Fe_3O_4} = \frac{0.1370 \times 2 \times 231.5/3 \times 159.7}{0.1666} \times 100\% = 79.47\%$$

Example 6-3

A chemically pure $AlPO_4$ sample can give 0.1126g of $Mg_2P_2O_7$. How many Al_2O_3 can be got?

Solution

$M(Mg_2P_2O_7)=222.6$ g·mol^{-1}; $M(Al_2O_3)=102.0$ g·mol^{-1}

$$Mg_2P_2O_7 — 2P — 2Al — Al_2O_3$$

$$m_{Al_2O_3} = m_{Mg_2P_2O_7} \frac{M(Al_2O_3)}{M(MgP_2O_7)}$$

So
$$m_{Al_2O_3} = (0.1126 \times 102.0/222.6) \text{g} = 0.05160 \text{g}$$

Example 6-4

Ammonium ion can be precipitated into $(NH_4)_2PtCl_6$ with H_2PtCl_6 and ignited to metal Pt. The equation is as follows:

$$(NH_4)_2PtCl_6 \longrightarrow Pt + 2NH_4Cl + 2Cl_2 \uparrow$$

If 0.1032g of Pt is got, calculate the mass of NH_3 in the sample.

Solution

$$M(NH_3) = 17.03 \text{g} \cdot \text{mol}^{-1}; \quad M(Pt) = 195.1 \text{g} \cdot \text{mol}^{-1}$$

$$(NH_4)_2PtCl_6 - Pt - 2NH_3$$

$$m_{NH_3} = m_{Pt} \frac{2M(NH_3)}{M(Pt)}$$

So

$$m_{NH_3} = (0.1032 \times 2 \times 17.03 / 195.1) \text{g} = 0.01802 \text{g}$$

Section 5 Precipitation Titrations

第 5 节 沉淀滴定法

Fig.6-13 Shows a typical precipitation titration, using the analysis of Ag^+ in an aqueous solution by its titration with Cl^- as an example. An insoluble precipitate [AgCl(s), in this case] is formed as the titrant is added to the sample, The dashed lines shown in the titration curve indicate the volume of titrant that is needed to reach the equivalence point and gives the concentration of Ag^+ this point by using the function pAg, where $pAg=-lg[Ag^+]$ ($pX=-lg[X^-]$). The titration curve shown is the response expected for a 10.00 mL aliquot of a 0.01000 mol $\cdot$ L^{-1} aqueous solution of Ag^+ that is titrated using 0.0050 mol $\cdot$ L^{-1} Cl^-.

The formation of a precipitate can be used as the basis of a titration, provided that there is a suitable way of determining when a stoichiometric amount of titrant has been added. In this section we briefly introduce argenometry (after the Latin word argenometry for "silver"), a type of the precipitation titrations.

图 6-13 为典型的沉淀滴定装置及滴定曲线，以 Cl^- 滴定分析水溶液中的 Ag^+ 为例。将滴定剂滴加到样品中形成不溶性沉淀［在这种情况下为 AgCl(s)］。滴定曲线中的虚线表示到达化学计量点时所需的滴定剂的体积及此时 pAg，通过 $pAg=-lg[Ag^+]$ 计算出 Ag^+ 的浓度（本书中 $pX=-lg[X^-]$）。该曲线为 0.0050mol $\cdot$ L^{-1} 的 Cl^- 滴定 10.00mL 0.01000mol $\cdot$ L^{-1} Ag^+ 的滴定曲线。

沉淀的形成是沉淀滴定的基础，沉淀滴定的前提是要用一种合适的方法确定滴定剂的化学计量点。在本节中，我们主要介绍银量法，它是一种常用的沉淀滴定法。

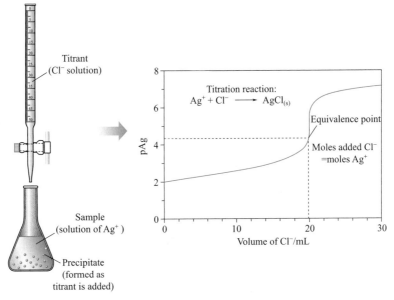

Fig.6-13 A typical precipitation titration curve

图 6-13 典型的沉淀滴定装置及滴定曲线

1. Precipitation Reactions

The titrations of chloride, bromide, iodide and other anions with silver(I) are based on the formation of insoluble silver salts, such as

$$Ag^+ + Cl^- \longrightarrow AgCl(s)$$
$$Ag^+ + SCN^- \longrightarrow AgSCN(s)$$

These procedures are termed **argenometry**.
Take the formation of AgCl precipitate as an example. Its solubility product constant is

1. 沉淀滴定反应

氯化物、溴化物、碘化物和其他阴离子与一价银的滴定都是建立在生成不溶性银盐反应的基础上,这种分析方法统称为银量法。以氯化银沉淀为例,其溶度积常数表达式为式(6.5)。

$$K_{sp} = a(Ag^+)a(Cl^-) = [Ag^+]\gamma(Ag^+)[Cl^-]\gamma(Cl^-) \tag{6.5}$$

The solubility of AgCl precipitate is very small, so $\gamma(Ag^+)$ and $\gamma(Cl^-)$ are all approximate to 1, and the eq.(6.5) can be simplified to

由于氯化银沉淀物的溶解度非常小,因此 $\gamma(Ag^+)$ 和 $\gamma(Cl^-)$ 均接近 1,故可将式(6.5)简化为式(6.6)。

$$K_{sp}(AgCl) = [Ag^+][Cl^-] \tag{6.6}$$

Then, we have

$$[Ag^+] = [Cl^-] = \sqrt{K_{sp}(AgCl)} \tag{6.7}$$

For the insoluble salt AB, if we take the side reactions into account, then

对于不溶性盐 AB,如果考虑副反应作用,可得式(6.8)。

$$\frac{[A']}{[A]} = \alpha_A, \frac{[B']}{[B]} = \alpha_B \tag{6.8}$$

Where α_A and α_B are the corresponding side reactions coefficient; [A'] and [B'] represent the total concentrations of A and B in the solution, respectively; [A] and [B] represent the free concentrations. So, the conditional solubility product is introduced

$$K'_{sp} = [A'][B'] = K_{sp}\alpha_A\alpha_B \tag{6.9}$$

2. Precipitation Reactions

Consider the titration of 0.1 mol·L⁻¹ KCl with 0.1 mol·L⁻¹ AgNO₃ solution. A titration curve can be prepared by plotting pCl ($-\lg[Cl^-]$) vs the volume of AgNO₃. At the beginning of the titration, we have 0.1 mol·L⁻¹ Cl⁻, and pCl is 1. As the titration continues, part of the Cl⁻ is removed from solution by precipitation as AgCl, and the pCl is determined by the concentration of the remaining Cl⁻; the contribution of Cl⁻ from dissociation of the precipitate is negligible, except near the equivalence point.

At the equivalence point, we have a saturated solution of AgCl.

Since $[Ag^+][Cl^-]=K_{sp}$, and $[Cl^-]=\sqrt{K_{sp}}=10^{-5}$ mol·L⁻¹, then pCl=5. After the equivalence point, there is excess Ag⁺ and the Cl⁻ concentration is determined from the concentration of Ag⁺ and K_{sp}, that is

Fig.6-14 shows the complete titration curve and the effect of reactant concentrations on the titration. Fig.6-15 illustrates how K_{sp} affects the titration of halide ions. The

2. 沉淀曲线

假设用 0.1mol·L⁻¹ AgNO₃ 滴定 0.1mol·L⁻¹ KCl。通过绘制 pCl 对加入 AgNO₃ 体积的曲线，可以得到滴定曲线。在滴定开始时，由于 [Cl⁻]=0.1mol·L⁻¹，pCl=1。随着滴定的进行，部分 Cl⁻ 从溶液中沉淀出来，pCl 由剩余 Cl⁻ 的浓度决定；除了接近化学计量点外，沉淀的解离都可以忽略不计。

化学计量点时，氯化银为饱和溶液。由于 $[Ag^+][Cl^-]=K_{sp}$，则 [Cl⁻]=10⁻⁵mol·L⁻¹，即 pCl=5。在化学计量点后，Ag⁺ 过量，Cl⁻ 浓度由 Ag⁺ 浓度和 K_{sp} 确定。

图 6-14 显示了完整的滴定曲线和滴定剂浓度对滴定的影响。图 6-15 说明了 K_{sp} 对滴定的影响，溶度积最小的

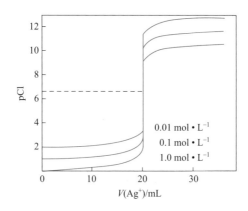

Fig.6-14 Titration curves showing the effect of concentration

图 6-14 浓度对滴定曲线的影响

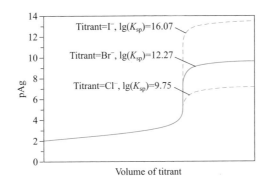

Fig.6-15　Titration curves showing the effect of K_{sp}

图 6-15　K_{sp} 对滴定曲线的影响

least solubility product, AgI, gives the sharpest change at the equivalence point. The larger the equilibrium constant for any titration reaction, the more pronounced the change will be in concentration near the equivalence point.

碘化银在化学计量点处变化最大。滴定反应的平衡常数越大，化学计量点附近的浓度变化就越明显。

$$[Cl^-] = \frac{K_{sp}}{[Ag^+]} \tag{6.10}$$

3. General Approach to Calculations

3. 通用计算过程

Just as we have seen complexometric titrations and acid-base titrations, we can also predict the shape of a curve for a precipitation titration. The response we get will be similar to what we see for a complexometric titration, in that we will again be plotting pM versus the volume of titrant. However, the value we calculate for pM will now depend on our precipitaion reaction rather than on complex formation (fig 6-16). We can again break this titration into four general regions: (1) the beginning of the titration, (2) before the equivalence point, (3) at the equivalence point, and (4) after the equivalence point.

正如配位滴定和酸碱滴定一样，沉淀滴定曲线即绘制 pM 与滴定剂体积的关系曲线（见图 6-16）。只不过 pM 值取决于沉淀反应，而不是配合物的形成。可将滴定分为四个区域：（1）滴定开始时，（2）化学计量点前，（3）化学计量点处，（4）化学计量点后。

(1) The beginning of the titration　At the beginning of the titration, the value of pM is simply obtained from the concentration of the analyte in the original sample (c_M). Original sample (0% titration):

（1）滴定开始前　滴定开始时，pM 的值可以从原始样品中待测物质的浓度（c_M）获得。

$$pM = -\lg c_M \tag{6.11}$$

(2) Before the equivalence point　Once we have added the titrant, it will react with the analyte to form a precipitate. Because we are generally using a precipitation reaction

（2）化学计量点前　滴定剂加入后，滴定剂会与待测物质反应形成沉淀。滴定过程中，由于沉淀反应的发生，得到

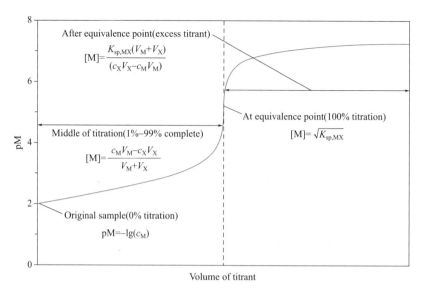

Fig.6-16 Equations for predicting the response for the titration of a metal ion (M) with a precipitating agent (X) to form a 1∶1 precipitate

图 6-16 金属离子（M）与沉淀剂（X）滴定形成 1∶1 沉淀物质的计算公式

with a low solubility product for this type of titration, we can safely assume that this reaction uses essentially all of the added titrant until the titration is about 99% complete. This means we can determine the amount of analyte that remains in solution by simply subtracting the moles of added titrant ($c_X V_X$) from the original moles of analyte ($c_M V_M$) and dividing this difference by the total volume of the sample/titrant mixture ($V_M + V_X$). We can express this relationship by using equation (6.12) for the formation of a 1∶1 precipitate.

Middle of titration (1%-99% complete):

$$[M] = \frac{c_M V_M - c_X V_X}{V_M + V_X} \tag{6.12}$$

The value of pM is then estimated by using this estimated [M] for the sample/titrant mixture. Similar expressions, with coefficients for some of the preceding terms other than one, can be derived for the formation of precipitates that do not have a 1∶1 stoichiometry (e. g., MX_2).

(3) At the equivalence point　At the equivalence point, we have added exactly enough of X to precipitate all of M. In this case, the only soluble M or X in solution is that which is produced from the precipitate as an equilbrium is established in the sample/titrant mixture. The extent to

溶度积较小的产物。可以确定，反应过程中加入的滴定剂基本上都参与了反应，直到滴定程度达到约 99%。用分析物的物质的量（$c_M V_M$）减去加入滴定剂的物质的量（$c_X V_X$），差值除以样品与滴定剂混合物的总体积（$V_M + V_X$），即可计算出分析物在溶液中的含量。可以用式（6.12）来表示化学计量比为 1∶1 的沉淀形成关系。

然后，根据滴定过程计算出的 [M] 值计算 pM 的数值。对于一些化学计量系数不是 1，化学计量比不是 1∶1 的沉淀反应（例如 MX_2），可以推导出类似的表达式。

（3）化学计量点处　在化学计量点时，由于已加入足够的 X 使 M 沉淀完全。在这种情况下，溶液中溶解的 M 或 X 是由滴定过程中建立的沉淀溶解平衡产生的。MX 沉淀溶解形成 M 或 X

which M or X is formed from precipitate MX will be described by the solubility product for this reaction. For a 1∶1 precipitate, M and X will be formed in a 1∶1 ratio as some of MX redissolves. This means that [M] =[X] at the equivalence point, or $K_{sp,MX}$=[M][X]=[M]². We can now rearrange this expression and use it to find the value of [M] at this point in the titration curve.

Equivalence point for a 1∶1 precipitate (100% titration):

$$[M] = \sqrt{K_{sp,MX}}$$ (6.13)

Again, similar but more complex expressions can be derived for determining the value of [M] at the equivalence point for a precipitate that does not have a 1∶1 stoichiometry, as occurs for MX_2.

(4) After the equivalence point After the equivalence point for any precipitation titration, the concentration of X in solution will be approximately equal to the moles of excess added titrant divided by the total volume of the sample/titrant mixture, as given by $[X] = \dfrac{c_X V_X - c_M V_M}{V_M + V_X}$.

We can substitute this relationship for [X] in the expression to solve for [M], as shown below.

After Equivalence Point (Excess Titrant):

$$[M] = \frac{K_{sp,MX}(V_M + V_X)}{c_X V_X - c_M V_M}$$ (6.14)

4. End-point Detection

Three techniques are commonly employed to detect the end point in precipitation titrations: potentiometric methods, indicator methods and light-scattering methods. In this section we discuss three types of indicator methods applied to the titration of Cl^- with Ag^+.

(1) Mohr titration The Mohr titration for chloride was published more than a hundred years ago and is still used. A colored precipitate at the end point is formed in Mohr titration. In the Mohr titration, Cl^- is titrated with Ag^+ in the presence of CrO_4^{2-} (chromate, dissolved as $Na_2CrO_4^{2-}$).

的程度是由该反应的溶解积决定的。对于化学计量比是1∶1的沉淀反应,随着MX的部分再溶解,M和X将以1∶1的比例在溶液中产生。这意味着在化学计量点时,[M]=[X],或者$K_{sp,MX}$=[M][X]=[M]²。因此,可以重新整理得到式(6.13),并用其确定滴定曲线上对应的[M]值。

同样,可以推导出类似但更复杂的表达式,用于计算化学计量比不是1∶1的沉淀物(如MX_2)在化学计量点的[M]值。

(4)化学计量点后 在化学计量点后,溶液中X的浓度约等于过量加入滴定剂的物质的量除以样品与滴定剂混合液的总体积。将[X]带入溶度积表达式,即可求出[M]值,见式(6.14)。

4. 终点检测

沉淀滴定法终点检测的常用方法有三种:电位法、指示剂法和光散射法。在本节中,我们将讨论三种用Ag^+滴定Cl^-的检测方法。

(1)莫尔法 用于测定氯化物的莫尔法始于一百多年前,至今仍在使用。其是在CrO_4^{2-}存在下,用Ag^+滴定Cl^-,滴定终点形成有色沉淀。

titration reaction:
$$Ag^+ + Cl^- \longrightarrow AgCl(s) \downarrow$$
$$\text{(white)}$$

end-point reaction:
$$2Ag^+ + CrO_4^{2-} \longrightarrow Ag_2CrO_4(s) \downarrow$$
$$\text{(red)}$$

Determination of NaCl in NaOH
(Mohr titration)
烧碱中 NaCl 分析（莫尔法）

The AgCl precipitates before Ag_2CrO_4. The color of AgCl is white; dissolved CrO_4^{2-} is yellow; and Ag_2CrO_4 is red. The end point is indicated by the first appearance of red Ag_2CrO_4. Reasonable control of CrO_4^{2-} concentration is required for Ag_2CrO_4 precipitation to occur at the desired point in the titration.

At the equivalence point, the concentration of Ag^+ should be

$$[Ag^+] = [Cl^-] = \sqrt{K_{sp,AgCl}} = \sqrt{1.56 \times 10^{-10}} = 1.25 \times 10^{-5} (mol \cdot L^{-1})$$

So Ag_2CrO_4, should precipitate just when $[Ag^+] = 1.25 \times 10^{-5}$ mol·L^{-1}. The solubility constance of Ag_2CrO_4 is 9.0×10^{-12}. Then the concentration of CrO_4^{2-} should be

$$[CrO_4^{2-}] = \frac{K_{sp,Ag_2CrO_4}}{[Ag^+]^2} = \frac{9.0 \times 10^{-12}}{1.56 \times 10^{-10}} = 5.8 \times 10^{-2} (mol \cdot L^{-1})$$

If the concentration is greater than this, then Ag_2CrO_4 will begin to precipitate when $[Ag^+]$ is less than 1.25×10^{-5} mol·L^{-1} (before the equivalence point). If it is less than 5.8×10^{-2} mol·L^{-1}, then the $[Ag^+]$ will have to exceed 1.25×10^{-5} mol·L^{-1} (after the equivalence point) before the beginning of Ag_2CrO_4, precipitation. In practice, the indicator concentration is controlled between 2×10^{-3} mol·L^{-1} and 5×10^{-3} mol·L^{-1}.

Because some excess Ag_2CrO_4 is necessary for visual detection, the color is not seen until after the true equivalence point. We can correct for this titration error in two ways. One is by means of a blank titration with no chloride presence. The volume of Ag^+ needed to form a detectable red color is then subtracted from $V(Ag^+)$ in the Cl^- titration. Alternatively, we can standardize the

由于 AgCl 的溶解度小于 Ag_2CrO_4 的溶解度，AgCl 在 Ag_2CrO_4 之前析出。AgCl 的颜色为白色，CrO_4^{2-} 溶液为黄色，而 Ag_2CrO_4 的颜色是红色。当出现红色的 Ag_2CrO_4 时即指示到达滴定终点。为了使 Ag_2CrO_4 沉淀在滴定预期点出现，需要合理控制 CrO_4^{2-} 浓度。

在化学计量点时，Ag^+ 的浓度为：

所以，Ag_2CrO_4 应在 $[Ag^+] = 1.25 \times 10^{-5}$ mol·L^{-1} 时析出。由于 Ag_2CrO_4 的溶解度常数为 9.0×10^{-12}，此时，CrO_4^{2-} 的浓度为：

如果 CrO_4^{2-} 浓度大于该值，那么当 Ag^+ 浓度小于 1.25×10^{-5} mol·L^{-1}（在化学计量点之前）时，Ag_2CrO_4 就开始析出；如果 CrO_4^{2-} 浓度小于 5.8×10^{-2} mol·L^{-1}，则 Ag^+ 浓度必须超过 1.25×10^{-5} mol·L^{-1}（在化学计量点之后），Ag_2CrO_4 才开始沉淀。实际使用时，指示剂 CrO_4^{2-} 的浓度需控制在 $2 \times 10^{-3} \sim 5 \times 10^{-3}$ mol·L^{-1} 之间。

由于过量的 Ag_2CrO_4 是目测所必需的，所以在化学计量点之后才会看到颜色。可以通过两种方法来校正这种滴定误差。一种是在没有氯化物存在的情况下，进行空白实验，确定形成可检测的红色 Ag_2CrO_4 所需 Ag^+ 的体积，然后在 Cl^- 滴定时，从测得的 $V(Ag^+)$ 中将该

AgNO₃ by the Mohr titration, using a standard NaCl solution and titrant volumes similar to those for the titration of unknown.

For the Mohr titration, the pH should be controlled in the range 6.5-10.5. If the pH is too low, the concentration of CrO_4^{2-} is reduced by the equilibrium

$$2CrO_4^{2-} + 2H^+ \longrightarrow 2HCrO_4^- \longrightarrow Cr_2O_7^{2-} + H_2O$$

If the pH is too high, AgOH(s) may precipitate. About pH=8 is ideal for the titration; solid calcium carbonate will buffer the solution at about this pH. The Mohr method is useful for Cl^- and Br^-, but not for I^- or SCN^-.

(2) Volhard titration The Volhard titration is actually a procedure for the titration of Ag^+. To determine Cl^-, a back titration is necessary. First, the Cl^- is precipitated by a known, excess quantity of standard AgNO₃.

$$Ag^+ + Cl^- \longrightarrow AgCl(s)\downarrow$$

Instead of filtering out the silver chloride precipitate, it is faster to add nitrobenzene and shake vigorously. Nitrobenzene is a liquid that is immiscible with water and heavier than water. Shaking with nitrobenzene coagulates and coats the silver chloride precipitate, so that it cannot react with the aqueous layer of solution during the titration. The excess AgNO₃ stays in the aqueous layer, and then the excess Ag^+ is titrated with standard KSCN in the presence of Fe^{3+} [NH₄Fe(SO₄)₂, ferric alum, is the indicator].

Determination of NaCl in NaOH (Volhard titration)
烧碱中 NaCl 分析（佛尔哈德法）

$$Ag^+ + SCN^- \longrightarrow AgSCN(s)\downarrow$$

The titration with thiocyanate is carried out in acidic solution. When all the Ag^+ has been consumed, the SCN^- reacts with Fe^{3+} to form a red complex (The color change at the end point is not extremely sharp, but it can be detected with a little practice).

$$Fe^{3+} + SCN^- \longrightarrow FeSCN^{2+}$$
$$\text{(red)}$$

The appearance of the red color signals the end point. The amount of required SCN^- for the back titration tells us that of Ag^+ left over from the reaction with Cl^-. Because the total amount of Ag^+ is known, the amount consumed by Cl^- can then be calculated.

In the analysis of Cl^- by the Volhard titration, the end point slowly fades because AgCl is more soluble than AgSCN. The AgCl slowly dissolves and is replaced by AgSCN. To prevent this secondary reaction from happening, two techniques are commonly used. One is to filter off the AgCl and titrate only the Ag^+ in the filtrate. An easier procedure is to shake a few milliliters of nitrobenzene, $C_6H_5NO_2$ with the precipitated AgCl prior to the back titration. Nitrobenzene coats the AgCl and effectively isolates it from attack by SCN^-, Br^- and I^-, whose silver salts are more soluble than AgSCN, may be titrated by the Volhard titration without isolating the silver halide precipitate.

(3) Fajans titration　The Fajans titration uses an adsorption indicator. To see how this works, we must consider the electric charge of a precipitate. When Ag^+ is added to Cl^-, there will be excess Cl^- ions in solution prior to the equivalence point. Some Cl^- is selectively adsorbed on the AgCl surface, imparting a negative charge to the crystal surface. After the equivalence point, there is excess Ag^+ in solution. Adsorption of the Ag^+ cations on the crystal surface creates a positive charge on the particles of precipitate. The abrupt change from negative charge to positive charge occurs at the equivalence point.

Common adsorption indicators are anionic dyes, which are attracted to the positively charged particles produced immediately after the equivalence point. The adsorption of the negatively charged dye on the positively charged surface changes the color of the dye by interactions that are not well understood. The color change signals the end point in the titration. Because the indicator reacts with the

滴定过程中，红色的出现标志着终点的到达。由于加入的 Ag^+ 的总量是已知的，根据返滴定过程中加入的 SCN^- 的量，即可算出与 Cl^- 反应后剩余的 Ag^+ 的量，进而计算出 Cl^- 消耗的 Ag^+ 的量。

用佛尔哈德法测定 Cl^- 时，由于 AgCl 比 AgSCN 更易溶解，所以 AgCl 缓慢溶解并反应生成 AgSCN，使终点颜色逐渐消失。为了防止上述二次反应的发生，通常采用两种方法。一种是过滤除去 AgCl，只对滤液中的 Ag^+ 进行滴定；另一种更简单的方法是在返滴定之前，在 AgCl 沉淀中加入几毫升硝基苯（$C_6H_5NO_2$）并摇匀。由于硝基苯包覆在 AgCl 表面，可以有效地防止 AgCl 与 SCN^-、Br^- 和 I^- 发生反应（AgCl 的溶解度比 AgSCN 大），从而可以用佛尔哈德法进行滴定，而无需分离出卤化银沉淀。

（3）法扬司法　法扬斯法使用的是吸附指示剂，要了解其测定原理，我们必须考虑沉淀物表面所带的电荷。当将 Ag^+ 加入到含有 Cl^- 的溶液中时，在化学计量点前，溶液中存在过量的 Cl^-，由于一些 Cl^- 选择性地吸附在 AgCl 表面，使 AgCl 晶体表面带有负电荷；在化学计量点后，溶液中有过量的 Ag^+，由于 Ag^+ 吸附在 AgCl 晶体表面，使 AgCl 沉淀颗粒带有正电荷。从负电荷到正电荷的突变是在化学计量点处发生的。

Determination of NaCl in NaOH
烧碱中 NaCl 分析（法扬司法）

常见的吸附指示剂是阴离子染料，它们被化学计量点后立即产生的带正电荷的沉淀颗粒所吸附。带负电荷的染料

precipitate surface it is desirable to have as much surface area as possible. This means performing the titration under conditions that tend to keep the particles as small as possible, because small particles have more surface area than an equal volume of large particles. Low electrolyte concentration helps to prevent coagulation of the precipitate and maintain small particle size.

The indicator most commonly used for AgCl is dichlorofluorescein. This dye has a greenish yellow color in solution but runs pink when it is absorbed to AgCl. Because the indicator is a weak acid and must present in its anionic form, the pH of the reaction must be controlled. The dye eosin is useful in the titration of Br^-, I^- and SCN^-. It gives a sharper end point than dichlorofluorescein and is more sensitive. It cannot be used for AgCl because the eosin anion is more strongly bound to AgCl than Cl^- ion. Eosin will bind to the AgCl crystallites even before the particles become positively charged.

In all argentometric titrations, but especially with adsorption indicators, strong light (such as daylight through a window) should be avoided. Light causes decomposition of the sliver salts, and adsorbed indicators are especially light sensitive.

Experiment 6 Determination of Barium in $BaCl_2 \cdot 2H_2O$

1. Purpose of the experiment

(1) Determine the content of barium in the reagent of $BaCl_2 \cdot 2H_2O$.
(2) Master the basic operation techniques of gravimetric analysis such as precipitation, filtration, washing and burning.
(3) Further understand the precipitation theory of crystalline precipitation.

2. Experimental Principle

The content of barium in $BaCl_2 \cdot 2H_2O$ was determined by the following reaction:

$$Ba^{2+} + SO_4^{2-} \longrightarrow BaSO_4 \downarrow$$

$BaSO_4$ is a typical crystalline precipitate. When precipitates are initially formed, they are usually fine crystals, which can easily pass through the filter paper during filtration. Therefore, in order to obtain relatively pure and coarse $BaSO_4$ crystals, special attention should be paid to selecting the precipitation conditions conducive to the formation of coarse crystals during the precipitation of $BaSO_4$. The determination steps are summarized as follows:

$BaCl_2 \cdot 2H_2O$ sample $\xrightarrow{\text{Weighing}}$ $\xrightarrow{\text{Dissolving and diluting with water}}$ $\xrightarrow{\text{Add diluted HCl}}$ $\xrightarrow{\text{Heating to boiling}}$ $\xrightarrow{\text{Slowly add hot diluted } H_2SO_4 \text{ and stiring}}$ $\xrightarrow{\text{Aging}}$ $\xrightarrow{\text{Filtrating}}$ $\xrightarrow{\text{Move the precipitate to the crucible}}$ $\xrightarrow{\text{Drying}}$ $\xrightarrow{\text{Burning}}$ $\xrightarrow{\text{Cooling}}$ $\xrightarrow{\text{Weighing}}$ To a constant weight

When precipitation is separated out of the solution, the precipitation is contaminated due to its precipitation phenomenon. Anions such as NO_3^-, ClO_3^- and Cl^- are often co-precipitated in the form of barium salt, while cations such as alkali metal ions of Ca^{2+} and Fe^{3+} are often co-precipitated in the form of sulfate or hydrogen sulfate. As for which ions co-precipitate in the experiment and the magnitude of their influence, it depends on concentration of impurity ions and properties of the precipitation formed, such as solubility and dissociation. The addition of HCl solution, on the one hand, is to prevent co-precipitation of barium carbonate, barium phosphate, and barium hydroxide; on the other hand, to reduce the concentration of SO_4^{2-} in the solution, which is beneficial to obtain coarser crystal precipitation. In the determination of Ba^{2+}, dilute H_2SO_4 is used as precipitant. In order to precipitate $BaSO_4$ completely, excessive H_2SO_4 must be used. Since H_2SO_4 can be volatilized and removed when being burned at high temperature, H_2SO_4 brought by precipitation will not introduce errors. Therefore, the amount of precipitant can be excessive by 50% ~ 100%.

3. Appratus and Reagents

(1) Appratus   weighing bottle, beaker(150mL), 2 beakers(250mL), 2 beakers(400mL), two watch glasses (9cm), 2 graduated cylinders (10mL and 100mL) , 2 small test tubes, 2 glass rods, funnel holder, 2 long-necked funnels, 2 crucibles, crucible tongs, desiccator, 2 quantitative filter papers.

(2) Reagents $BaCl_2 \cdot 2H_2O$(solid), $2mol \cdot L^{-1}$ HCl solution, $2mol \cdot L^{-1}$ H_2SO_4 solution, $2mol \cdot L^{-1}$ HNO_3 solution, $0.01mol \cdot L^{-1}$ $AgNO_3$ solution.

4. Experimental Steps

This experiment does two simultaneous determinations.

(1) Weighing and dissolution of the sample   Take a dry and clean weighing bottle, weigh about 1.0g $BaCl_2 \cdot 2H_2O$ on the bench scale, and then accurately weigh it on the analytical balance. Add about half of the solid (0.4-0.6g) into a clean 250mL beaker (the beaker should be cleaned and dry,

mark beakers respectively), next, weigh the remaining solid and weighting bottle. The difference between two weights is weight of the sample that should be added into one beaker. Then take out of about 0.4 ~ 0.6g from the remaining solid into another beaker, weigh the remaining solid and weighing bottle, get weight of the second sample. Finally, dissolve with about 100mL of distilled water respectively.

(2) Precipitating Ba^{2+} with H_2SO_4

① Add 3mL 2mol·L^{-1} HCl solution to the first obtained solution, and then heat it with a small fire to near boiling (Do not make the solution boiling. Because the generated vapor may take away the droplets or cause the liquid to splash and cause the loss of solution.).

② Pour 3 ~ 5mL 2mol·L^{-1} H_2SO_4 solution into another 150mL beaker, dilute with 30mL of distilled water, and heat it to near boiling.

③ Use a dropper in your left hand to add the H_2SO_4 hot solution dropwise (approximately 2 ~ 3 drops per second at the beginning, and it can be slightly faster when more precipitates are deposited) into the barium chloride hot solution, while holding the glass rod in your right hand and stirring continuously. When stirring, do not allow the glass rod to touch the bottom or inner wall of beaker to avoid scratching and make the precipitate adhere to the inner wall of beaker, which is difficult to wash off. After only a few drops of H_2SO_4 are left, cover the beaker with a watch glass and let it stand for a few minutes.

④ When the precipitate is deposited on the bottom of beaker, add 1 ~ 2 drops of H_2SO_4 solution along the wall of beaker to check whether Ba^{2+} has precipitated completely. If there is turbidity in the supernatant, H_2SO_4 solution must be added until the precipitation is complete, and then cover the beaker with a watch glass (don't take out the glass rod, why?). Take the second solution and proceed with precipitation according to the above steps. After precipitation is completed, it is placed and aged for no less than 12h until the next experiment

(3) Burning and constant weight of empty crucible   Take two clean, dry crucibles and burn them in a constant temperature of high-temperature electric furnace. The burning temperature is 800-850℃. The first burning time is 30min, the crucible is cooled to room temperature (the time is half an hour), and then quickly weigh, and repeat burning for 15-20 min, cooling for half an hour, and then weighing until a constant weight.

(4) Filtration and washing of precipitation

① Filter. Take a dense ash-free filter paper, fold and put it in the funnel to form a water column Place the funnel on the funnel rack and place a clean 400mL beaker under the funnel to collect the filtrate.

② Filter and wash by pouring method. Prepare 400mL washing solution (add 2mol·L^{-1} H_2SO_4 solution 2mL per 100mL water). Pour the supernatant of precipitation on the filter paper, and then wash the precipitation 3 times by pouring method. Use washing solution about 10mL each time (why wash with dilute H_2SO_4 solution?), and then quantitatively move the precipitation to the filter paper, continue to wash with a small amount of dilute H_2SO_4 solution for 7-8 times, and concentrate the precipitate on the bottom of filter paper cone.

Take a few drops of filtrate with a clean test tube, add two drops of dilute HNO_3 solution, a drop of $AgNO_3$ solution, and observe whether there is white AgCl turbidity. The precipitate must be washed until there is no Cl^- in the filtrate (why?).

(5) Burning and weighing of precipitation   Carefully take out of the filter paper containing the precipitate, wrap it, and put it into an empty crucible that has been burned to a constant weight. First, heat it on an electric furnace. After the filter paper is ashed, it is burned in a high temperature electric furnace at 800-850℃, and then cool and weigh until the weight is constant.

Precautions for burning $BaSO_4$ precipitation:

① Before the filter paper is ashed, temperature should not be too high, so as to prevent the precipitated particles from flying away with the flame.

② The air should be sufficient when the filter paper is ashed, otherwise the sulfate will be easily reduced by carbon of filter paper. The reaction is as follows:

$$BaSO_4 + 4C \longrightarrow BaS + 4CO \uparrow \qquad BaSO_4 + 2C \longrightarrow BaS + 2CO_2 \uparrow$$

If this phenomenon occurs, measurement result will be low.

③ Burning temperature should not be too high. If it exceeds 900℃, $BaSO_4$ will also be reduced by carbon. If it exceeds 950℃, part of $BaSO_4$ will be decomposed as follows:

$$BaSO_4 \longrightarrow BaO + SO_3 \uparrow$$

It must be pointed out that the same balance and the same box of weights should be used throughout the experiment (why?).

5. Recording and Processing of Experimental Data

The result of quality analysis is often expressed as the percentage of tested component in the sample. The content of barium in $BaCl_2 \cdot 2H_2O$ measured in this experiment can be calculated by the quality of $BaSO_4$ precipitation and $BaCl_2 \cdot 2H_2O$ sample.

Because the mass of Ba equals $M(Ba)/M(BaSO_4)$ multiplied by mass of $BaSO_4$, in the formula $M(Ba)/M(BaSO_4)$ is the ratio of measured component formula quantity to the formula quantity of weighing form, it is a constant, this ratio is called chemical factor or conversion factor.

The percentage content of Ba in the sample is:

$$w_{Ba} = \frac{\frac{M(Ba)}{M(BaSO_4)} \times \text{the mass of } BaSO_4 (g)}{\text{the mass of the sample(g)}} \times 100\%$$

6. Questions

(1) In this experiment, dilute H_2SO_4 solution is used as precipitant. Can Na_2SO_4 be used instead? Why?

(2) When precipitating $BaSO_4$, why do we have to add a small amount of dilute HCl solution to the barium salt solution?

(3) When the precipitation starts, why do we have to add the hot dilute H_2SO_4 solution drop by drop while stirring constantly?

(4) Why should a small amount of washing solution be used each time when washing precipitation, and washing times are more frequent? Why do we wait for a washing solution to flow out as far as possible before adding the next washing solution?

(5) If barium salt solution contains the same concentration of NO_3^- and Cl^-, which ion co-precipitates more with $BaSO_4$? Why?

(6) What is the basis for weighing 0.4 ~ 0.6g of $BaCl_2 \cdot 2H_2O$ sample in this experiment? What does it matter if you weigh more or less? How is the amount of the precipitant calculated? Why is it necessary to overdose?

(7) If $BaCl_2$ is used as the precipitant to determine SO_4^{2-}, choose a suitable detergent from the following reagents to wash the $BaSO_4$ precipitate: ① H_2O; ② $BaCl_2$; ③ H_2SO_4; ④ HNO_3.

实验 6 $BaCl_2 \cdot 2H_2O$ 中钡的测定

1. 实验目的

（1）测定试剂 $BaCl_2 \cdot 2H_2O$ 中钡的含量；

（2）掌握沉淀、过滤、洗涤及灼烧等重量分析基本操作技术；

（3）加深理解晶形沉淀的沉淀理论。

2. 实验原理

测定 $BaCl_2 \cdot 2H_2O$ 中钡的含量，利用下式反应：$Ba^{2+} + SO_4^{2-} \longrightarrow BaSO_4\downarrow$。$BaSO_4$ 是典型的晶形沉淀。沉淀初生成时，常是细小的晶体，在过滤时易透过滤纸。因此，为了得到比较纯净而较粗大的 $BaSO_4$ 晶体，在沉淀 $BaSO_4$ 时，应特别注意选择有利于形成粗大晶体的沉淀条件。测定步骤概括如下：

$BaCl_2 \cdot 2H_2O$ 试样 $\xrightarrow{称量}$ 加水溶解、稀释 $\xrightarrow{}$ 加稀HCl $\xrightarrow{}$ 加热近沸 $\xrightarrow{}$ 缓慢加入热的稀H_2SO_4并不断搅拌 $\xrightarrow{}$ 陈化 $\xrightarrow{}$ 过滤 $\xrightarrow{}$ 将沉淀定量转移至坩埚 $\xrightarrow{}$ 干燥 $\xrightarrow{}$ 灼烧 $\xrightarrow{}$ 冷却 $\xrightarrow{称量}$ 直至恒重

当沉淀从溶液中析出时，由于其沉淀现象使沉淀沾污，如 NO_3^-、ClO_3^- 和 Cl^- 等阴离子常以钡盐的形式共沉淀，而碱金属离子 Ca^{2+} 和 Fe^{3+} 等阳离子常以硫酸盐或硫酸氢盐的形式共沉淀。至于在实验中哪些离子共沉淀及其影响的大小，取决于杂质离子的浓度及其所形成沉淀的性质，如溶解度、离解度等。

加入 HCl 溶液，一方面为了防止产生碳酸钡、磷酸钡、氢氧化钡等共沉淀；另一方面降低溶液中 SO_4^{2-} 的浓度，有利于获得较粗大的晶形沉淀。测定 Ba^{2+} 时，选用稀 H_2SO_4 作沉淀剂。为了使 $BaSO_4$ 沉淀完全，H_2SO_4 必须过量。由于高温灼烧时 H_2SO_4 可挥发除去，沉淀带入的 H_2SO_4 不致引入误差。因此，沉淀剂用量可过量 50% ~ 100%。

3. 仪器与试剂

（1）仪器 称量瓶1个、150mL烧杯1个、250mL和400mL烧杯各2个、9cm表面皿2块、10mL和100mL量筒各1个、小试管2个、玻璃棒2根、漏斗架、长颈漏斗2个、坩埚2个、坩埚钳、干燥器、定量滤纸2张。

（2）试剂 $BaCl_2 \cdot 2H_2O$（固体）、$2mol \cdot L^{-1}$ HCl溶液、$2mol \cdot L^{-1}$ H_2SO_4 溶液、$2mol \cdot L^{-1}$ HNO_3 溶液、$0.01mol \cdot L^{-1}$ $AgNO_3$ 溶液。

4. 实验步骤

（1）试样的称取及溶解 取一干燥洁净的称量瓶，在台称上称取约 1.0g $BaCl_2 \cdot 2H_2O$，再在分析天平上准确称量。将约一半（0.4～0.6g）的固体，倒入洁净的250mL烧杯（烧杯应洗涤到内壁不挂水珠，并在烧杯上分别编号）中，再称量剩余的固体及称量瓶重，两次重量之差，即为倒入烧杯中试样的重量。然后，从剩余的固体中再倒出约0.4～0.6g至另一烧杯中，称量剩余的固体及称量瓶重，即得第二份试样的重量。分别用约100mL蒸馏水溶解。

（2）用 H_2SO_4 沉淀 Ba^{2+}

① 在所得的第一份溶液中加入 3mL $2mol \cdot L^{-1}$ HCl溶液，用小火加热至近沸。注意不能使溶液沸腾，因为产生的蒸气可能把液滴带走或引起液体飞溅而使溶液损失。

② 在另一个150mL烧杯中，加入 3～5mL $2mol \cdot L^{-1}$ H_2SO_4 溶液，用30mL蒸馏水稀释，加热近沸。

③ 左手用滴管将 H_2SO_4 热溶液逐滴地（开始大约每秒钟加入2～3滴，待有较多沉淀析出时可稍快些）加入氯化钡热溶液中，同时右手持玻璃棒不断地搅拌。搅拌时，玻璃棒不要碰烧杯底或内壁以免划损烧杯，且使沉淀粘附在烧杯壁上，难以洗下。待只剩下数滴 H_2SO_4 后，用表面皿将烧杯盖好，静置数分钟。

④ 当沉淀沉积于烧杯底时，沿烧杯壁加入 1～2 滴 H_2SO_4 溶液，检验 Ba^{2+} 是否沉淀完全。如果上层清液中有浑浊出现，必须再加入 H_2SO_4 溶液，直到沉淀完全为止，然后将烧杯用表面皿盖好（不要取出玻璃棒，为什么？）。

取第二份溶液，按上述步骤进行沉淀。沉淀完毕后，放置陈化到下次实验，放置时间不少于12h。

（3）空坩埚的灼烧和恒重 取两个洁净、干燥的坩埚放入已恒温的高温电炉中灼烧，灼烧温度为800～850℃。第一次灼烧时间为30min，坩埚冷却至室温（时间为半小时），然后迅速进行称量，重复灼烧15～20min，冷却半小时，再称量，直到恒重。

（4）沉淀的过滤和洗涤

① 过滤装置。取一张致密的无灰滤纸，折叠好放在漏斗中并形成"水柱"。将漏斗放在漏斗架上，漏斗下放一洁净的400mL烧杯承接滤液。

② 用倾注法过滤和洗涤。配制400mL洗涤液（每100mL水中加入 $2mol \cdot L^{-1}$ H_2SO_4 溶液2mL）。先将沉淀上层清液倾注在滤纸上，再用倾注法洗涤沉淀3次。每次用洗涤液约10mL（为什么用稀的 H_2SO_4 溶液洗涤？），然后把沉淀定量转移到滤纸上，继续用少量稀 H_2SO_4 溶液洗涤7～8次，并使沉淀集中在滤纸圆锥体的底部。

用洁净的试管接取滤液数滴，加 2 滴稀 HNO_3 溶液，1 滴 $AgNO_3$ 溶液，观察是否有白色 AgCl 浑浊出现。沉淀必须洗涤到滤液中不含 Cl^- 为止（为什么？）。

（5）沉淀的灼烧和称量　小心取出装有沉淀的滤纸，包好后放入已灼烧到恒重的空坩埚内。先在电炉上加热，待滤纸灰化后在 800～850℃ 的高温电炉内灼烧，然后冷却、称量、直至恒重。

灼烧 $BaSO_4$ 沉淀时的注意事项：
① 在滤纸未灰化前，温度不要太高，以免沉淀颗粒随火焰飞散；
② 滤纸灰化时，空气要充足，否则硫酸盐易被滤纸上的碳还原。反应如下：

$$BaSO_4 + 4C \longrightarrow BaS + 4CO \uparrow \qquad BaSO_4 + 2C \longrightarrow BaS + 2CO_2 \uparrow$$

如果发生这种现象，将使测定结果偏低。
③ 灼烧温度不能太高，如超过 900℃，$BaSO_4$ 也会被碳还原。如超过 950℃，部分 $BaSO_4$ 将按下式分解：$BaSO_4 \longrightarrow BaO + SO_3 \uparrow$

必须指出，在整个实验过程中，应使用同一台天平和同一盒砝码（为什么？）。

5. 实验数据记录与处理

质量分析的结果常以试样中被测组分的百分含量来表示。本实验所测 $BaCl_2 \cdot 2H_2O$ 中钡的含量，可根据所得 $BaSO_4$ 沉淀的质量和试样 $BaCl_2 \cdot 2H_2O$ 质量来计算。Ba 的质量等于 $M(Ba)/M(BaSO_4)$ 乘以 $BaSO_4$ 的质量，式中 $M(Ba)/M(BaSO_4)$ 是待测组分的式量与称量形式的式量之比，它是一个常数，这一比值称为化学因数或换算因数。

试样中 Ba 的含量为：

$$w_{Ba} = \frac{\dfrac{M(Ba)}{M(BaSO_4)} \times BaSO_4 \text{的质量}(g)}{\text{试样的质量}(g)} \times 100\%$$

6. 思考题

（1）本实验用稀 H_2SO_4 溶液作沉淀剂，能否改用 Na_2SO_4？为什么？
（2）沉淀 $BaSO_4$ 时，为什么要在钡盐溶液中加少量稀 HCl 溶液？
（3）开始沉淀时，为什么要逐滴加入热的稀 H_2SO_4 溶液，还要不断搅拌？
（4）为什么洗涤沉淀时，每次用少量洗涤液，而且洗涤的次数要多？为什么要等有一份洗涤液尽量流出后才加入下一份洗涤液？
（5）如果钡盐溶液中含有相同浓度的 NO_3^- 和 Cl^-，哪一种离子和 $BaSO_4$ 共沉淀较多？为什么？
（6）本实验 $BaCl_2 \cdot 2H_2O$ 试样称取 0.4～0.6g 是根据什么？如果称取更多或更少有什么关系？沉淀剂的用量是怎样计算的？为什么要稍过量？
（7）如果以 $BaCl_2$ 为沉淀剂测定 SO_4^{2-}，从以下试剂中选择合适的洗涤剂洗涤 $BaSO_4$ 沉淀：① H_2O；② $BaCl_2$；③ H_2SO_4；④ HNO_3。

Experiment 7 Preparation and Calibration of Silver Nitrate Standard Solution, Determination of Chlorine Content in Tap Water (Mohr Method)

1. Purpose of the experiment

(1) Master preparation and calibration method of $AgNO_3$ standard solution.
(2) Master determination principle and method of Mohr Method.

2. Experimental principle

Mohr Method is often used to determine content of chlorine in some soluble chloride or tap water. This method is carried out in neutral or weakly alkaline solution, K_2CrO_4 solution is used as indicator and titrated with $AgNO_3$ standard solution. Because the solubility of AgCl is lower than that of Ag_2CrO_4, when AgCl is quantitatively precipitated, the excess $AgNO_3$ solution will form Ag_2CrO_4 precipitation with CrO_4^{2-}, indicating the arrival of the end point. The reaction formula is as follows:

$$Ag^+ + Cl^- \longrightarrow AgCl \downarrow \text{ (white)} \quad 2Ag^+ + CrO_4^{2-} \longrightarrow Ag_2CrO_4 \downarrow \text{ (Brick red)}$$

Titration must be carried out in a neutral or weakly alkaline solution, and the most suitable pH range is 6.5-10.5. If acidity is too high, no Ag_2CrO_4 precipitation will be produced, and if it is too low, Ag_2O precipitation will be formed.

Dosage of the indicator has an influence on accurate judgment of the titration end point, generally 5×10^{-3} mol·L^{-1} is appropriate.

3. Apparatus and reagents

(1) Apparatus   a set of titration analysis instrument.
(2) Reagents $AgNO_3$ solid, NaCl reference substance, 5% K_2CrO_4 solution.

4. Experimental steps

(1) Preparation and calibration of 0.01mol·L^{-1} $AgNO_3$ solution   The $AgNO_3$ standard solution can be directly prepared with dry reference $AgNO_3$, but the calibration method is generally adopted. The most commonly used reference substance for calibration of $AgNO_3$ solution is NaCl.

Preparation of $AgNO_3$ solution

Weigh $AgNO_3$_____g, dissolve in 250mL water, shake well and store in a brown reagent bottle with a glass stopper.

Accurately weigh_____g dried reference reagent NaCl in a small beaker, dissolve and move quantitatively to a 200mL volumetric flask, and dilute to the mark.

Take three 20.00mL portions of this solution, place them in 250mL conical flask, add 25mL water, add 1mL 5% K_2CrO_4 solution After fully shaking, titrate with $AgNO_3$ solution until the solution turns slightly brick red, which is the end point. Record the volume of $AgNO_3$ solution, and simultaneously measure it for 3 times. The exact concentration of $AgNO_3$ solution is calculated according to mass of NaCl and volume of AgNO₃ solution.

(2) Determination of chlorine content in tap water　　Take accurately three samples of 100.00mL tap water pour them in 250mL conical flask respectively, and add 1mL 5% K_2CrO_4 indicator. After fully shaking, titrate with $AgNO_3$ standard solution until the solution turns brick red, which is the end point. Simultaneously measure it for 3 times. The chlorine content in tap water samples is calculated according to concentration of $AgNO_3$ standard solution and volume that used for titration.

Notes:

① The distilled water that is used to prepare $AgNO_3$ solution should not contain chloride ions. The prepared $AgNO_3$ solution should be stored in a brown bottle, and titrate with a brown acid burette.

② If pH is over 10.5, Ag_2O precipitation is produced. If pH is less than 6.5, most of CrO_4^{2-} is transformed into $Cr_2O_7^{2-}$, which delays appearance of the end point. If there is ammonium salt, in order to avoid production of $Ag(NH_3)_2^+$, pH of the solution during titration should be controlled within the range of 6.5-7.0. When concentration of NH_4^+ is more than $0.1\text{mol} \cdot L^{-1}$, Mohr Method cannot be used for determination.

5. Recording and processing of experimental data

(1) Calculation of accurate concentration of $0.01\text{mol} \cdot L^{-1}$ $AgNO_3$ solution.

(2) Calculation of Cl^- content in tap water.

6. Questions

(1) In titration, what range should acidity of test solution be controlled? Why? When there is NH_4^+, what is the acidity that should control? Why should it be different?

(2) Should dosage of K_2CrO_4 indicator be controlled during titration? Why?

(3) Why should the solution be fully shaken during titration? If not, what is the impact on the determination result?

实验 7　硝酸银标准溶液的配制与标定、自来水中氯含量的测定
（莫尔法）

1. 实验目的

（1）掌握 $AgNO_3$ 标准溶液的配制及标定方法；

（2）掌握莫尔法的测定原理及方法。

2. 实验原理

某些可溶性氯化物或自来水中氯含量的测定常采用莫尔法。此方法是在中性或弱碱性溶液中，以 K_2CrO_4 溶液为指示剂，用 $AgNO_3$ 标准溶液进行滴定。由于 AgCl 的溶解度比 Ag_2CrO_4 的溶解度小，因此，溶液中首先析出 AgCl 沉淀，当 AgCl 定量沉淀后，过量的

AgNO₃ 溶液即与 CrO₄²⁻ 生成 Ag₂CrO₄ 沉淀，指示终点的到达。反应式如下：

$$Ag^+ + Cl^- \longrightarrow AgCl \downarrow（白色）\quad 2Ag^+ + CrO_4^{2-} \longrightarrow Ag_2CrO_4 \downarrow（砖红色）$$

滴定必须在中性或弱碱性溶液中进行，最适宜的 pH 值范围为 6.5～10.5。酸度过高，不产生 Ag₂CrO₄ 沉淀，过低则形成 Ag₂O 沉淀。指示剂的用量对滴定终点的准确判断有影响，一般以 $5×10^{-3}$ mol·L⁻¹ 为宜。

3. 仪器与试剂

（1）仪器　滴定分析仪器 1 套。

（2）试剂　AgNO₃ 固体、NaCl 基准物质、5% K₂CrO₄ 溶液。

4. 实验步骤

（1）0.01mol·L⁻¹ AgNO₃ 溶液的配制与标定　AgNO₃ 标准溶液可以直接用干燥的基准 AgNO₃ 来配制，但一般采用标定法。标定 AgNO₃ 溶液最常用的基准物质是 NaCl。

硝酸银标准溶液的制备

称取 AgNO₃_____g，溶于 250mL 水，摇匀后储存于带玻璃塞的棕色试剂瓶中。准确称取_____g 烘干后的基准试剂 NaCl 于小烧杯中，溶解后定量转移到 200mL 容量瓶中，稀释至刻度。

取 3 份 20.00mL 的此溶液，分别置于 250mL 锥形瓶中，加水 25mL，加 5% K₂CrO₄ 溶液 1mL，在充分摇动下，用 AgNO₃ 滴定至溶液呈微砖红色即为终点，记下 AgNO₃ 溶液的体积，平行测定 3 次。

根据 NaCl 的质量和 AgNO₃ 溶液的体积，计算 AgNO₃ 溶液的准确浓度。

（2）自来水中氯含量的测定　准确移取 100.00mL 自来水试样 3 份，分别置于 250mL 锥形瓶中，加 5% K₂CrO₄ 指示剂 1mL，在充分摇动下用 AgNO₃ 标准溶液滴定至溶液呈砖红色，即为终点。平行测定 3 次。根据 AgNO₃ 标准溶液的浓度和滴定消耗的体积，计算自来水样品中氯的含量。

注：

① 配制 AgNO₃ 溶液用的蒸馏水，不能含有氯离子。配好的 AgNO₃ 溶液应储存于棕色瓶中，滴定时使用棕色酸式滴定管。

② 如果 pH＞10.5，产生 Ag₂O 沉淀；pH<6.5 时，则大部分 CrO₄²⁻ 转变成 Cr₂O₇²⁻，使终点推迟出现。如果有铵盐存在，为了避免产生 Ag(NH₃)₂⁺，滴定时溶液的 pH 值应控制在 6.5～7.0 的范围内，当 NH₄⁺ 的浓度大于 0.1mol·L⁻¹ 时，便不能用莫尔法进行测定。

5. 实验数据记录与处理

（1）0.01mol·L⁻¹ AgNO₃ 溶液准确浓度的计算。

（2）自来水中 Cl⁻ 含量的计算。

6. 思考题

（1）滴定中，试液的酸度宜控制在什么范围？为什么？有 NH₄⁺ 存在时，在酸度控制上为什么要有所不同？

（2）滴定中对 K₂CrO₄ 指示剂的用量是否要控制？为什么？

（3）在滴定过程中为什么要充分摇动溶液？如果不充分摇动，对测定结果有何影响？

Exercises

6-1 Which of the following statements does not belong to the precipitation method requirements for the form of precipitation? ( ).
A. The solubility of the precipitate is small.
B. The precipitate is pure.
C. The precipitated particles are easy to filter and wash.
D. The molar mass of the precipitate is large.

6-2 Which of the following statements about the basic concept of gravimetric analysis is false? ( ).
A. The gasification method is a method of analyzing the weight loss of the sample.
B. The gasification method is suitable for the determination of volatile substances and moisture.
C. The basic data of the gravimetric method are all obtained by weighing.
D. The systematic error of gravimetric analysis is only related to the weighing error of the balance.

6-3 The ashless filter paper used in the gravimetric analysis refers to the ash content of each filter paper, it should be ( ).
A. no weight   B. <0.2mg   C. >0.2mg   D. =0.2mg

6-4 In the gravimetric determination of SO_4^{2-}, the $BaSO_4$ precipitation is ( ).
A. non-crystalline precipitation   B. crystalline precipitation
C. colloid   D. amorphous precipitation

6-5 When using SO_4^{2-} to precipitate Ba^{2+}, adding excess SO_4^{2-} can make Ba^{2+} precipitation more complete, which is the use of ( ).
A. complexing effect   B. common ion effect
C. salt effect   D. acidic effect

6-6 Which of the following statements is true? ( )
A. The weighing form and precipitation form should be the same.
B. The weighing form and precipitation form must be different.
C. The weighing form and precipitation form can be different.
D. Neither the weighing form nor precipitation form can contain water molecules.

6-7 The precipitation conditions for crystal precipitation are ( ).
A. thick, cold, slow, stiring, aged   B. thin, hot, fast, stiring, aged
C. thin, hot, slow, stiring, aged   D. thin, cold, slow, stiring, aged

6-8 Which of the following statements about burning containers is false? ( )
A. The firing container must have a constant weight before or after firing the sediment.
B. The constant weight should be burned at least twice, and only when the two weights are consistent that can be considered as a constant weight.
C. When weighing after burning, the cooling time should be the same, and the constant weight is valid.
D. The time for burning the glass sand filter can be shorter.

6-9 For the determination of NaCl in soda ash by Mohr titration, the indicator selected is ( ).
A. $K_2Cr_2O_7$ B. K_2CrO_4 C. KNO_3 D. $KClO_3$

6-10 The indicator used by Fajans titration is (   ).
    A. potassium chromate   B. ferric alum
    C. adsorption indicator   D. self indicator

6-11 When the Volhard titration is used to determine the Ag^+ content in water, the color of the end point is (   ).
    A. red   B. pure blue   C. yellow-green   D. blue-violet

6-12 When titrating Ag^+ with ammonium thiocyanate standard solution using ferric ammonium alum as indicator, which of the following conditions should be carried out ?(   )
    A. Acidic   B. Weak acid   C. Alkaline   D. Weak base

6-13 When the Mohr titration is used to determine the content of Cl^-, the pH value of the medium is required to be in the range of 6.5 to 10.5. If the acidity is too high, (   ).
    A. AgCl precipitation is incomplete.   B. Ag_2O precipitation is formed.
    C. The ability of AgCl precipitation to adsorb Cl^- is enhanced.
    D. Ag_2CrO_4 precipitation is not easy to form.

6-14 Mohr titration in precipitation titration refers to (   )
    A. Argentometry with potassium chromate as indicator.
    B. an analytical method for titrating Ba^{2+} in the test solution with K_2CrO_4 standard solution with $AgNO_3$ as indicator.
    C. Argentometry using an adsorption indicator to indicate the end point of the titration.
    D. Argentometry with ferric ammonium alum as indicator.

6-15 The solubility products of AgCl and Ag_2CrO_4 are 1.8×10^{-10} and 1.2×10^{-12} respectively, which of the following statements is true? (   )
    A. AgCl has the same solubility as Ag_2CrO_4.
    B. The solubility of AgCl is greater than that of Ag_2CrO_4.
    C. The two types are different, and the solubility can not be judged directly by the solubility product.
    D. They are all insoluble salts, and their solubility is meaningless.

6-16 True or false
    (1) The precipitation form and the weighing form are the same sometimes but different sometimes.
    (2) The precipitation must be pure and free from precipitants and other impurities.
    (3) The weighing form of the precipitation should have as much molar mass as possible.
    (4) In gravimetric analysis, the solubility of the precipitation can be greatly reduced by the common ion effect.
    (5) The salt effect can make the solubility of the precipitation decrease with the increase of the electrolyte concentration in the solution.
    (6) The higher the concentration of the strong electrolyte, the higher the charge of its ions and precipitate crystallized ions, and the more serious the influence of the salt effect.
    (7) The acidity of the solution has no effect on the solubility of the precipitate.
    (8) The purity of precipitation can be improved by reducing concentration of easily-adsorbed impurity ions.

(9) K_{sp} is called the activity product and is related to not only temperature but also concentration.

(10) When Mohr titration is used to determine the Cl^- content, it should be carried out in a neutral or weakly acidic solution.

(11) The silver nitrate standard solution should be titrated in a brown basic burette.

(12) When the concentration of $AgNO_3$ solution is calibrated with sodium chloride reference reagent, the acidity of the solution is too large, which has no effect on the calibration result.

(13) Mohr titration for the determination of Cl^- in water uses the direct method.

(14) In the precipitation titration, the argentometry mainly refers to Mohr titration, Volhard titration and Fajans titration.

(15) According to the common ion effect, the more precipitant added, the more complete the precipitation.

(16) In Mohr titration, the amount of indicator added has no effect on the measurement results.

(17) When titrating Ag^+ with NH_4SCN standard solution with ferric ammonium alum as indicator, it should be carried out under alkaline conditions.

(18) Mohr titration uses ferric alum as an indicator, while Volhard titration uses potassium chromate as an indicator.

(19) In precipitation titration, various indicators indicating the end point is in a specific range of acidity.

(20) When the content of Cl^- is determined by Fajans titration, dichlorofluorescein ($K_a=1.0\times10^{-4}$) is used as the indicator, and pH value of the solution should be greater than 4 and less than 10.

6-17 What is the basic principle of gravimetric analysis?

6-18 In gravimetric analysis, what are the requirements for precipitation form and weighing form respectively?

6-19 What are the factors that affect the solubility of precipitates?

6-20 Why does pH value of the solution need to be controlled between 6.5 and 10.5 when determining Cl^- by Mohr titration?

6-21 Why should nitrobenzene be added before titration when Cl^- is determined by Volhard back titration?

6-22 It is known that at 25℃, the K_{sp} of AgCl=1.8×10^{-10}, what is the solubility of AgCl?

$(1.34\times10^{-5}\,mol\cdot L^{-1})$

6-23 The solution contains 1.0mol/L CO_3^{2-} and 0.1mol/L CrO_4^{2-}, when Ba^{2+} is added, which ion will precipitate first ($K_{sp,\,BaCO_3}=5.1\times10^{-9}$, $K_{sp,\,BaCrO_4}=1.2\times10^{-10}$)?

$(BaCrO_4)$

6-24 Weigh 0.5000g of iron-containing sample, after a series of treatments, its precipitation form is $Fe(OH)_3\cdot nH_2O$, weighing form is Fe_2O_3, and its mass is 0.4990g. Calculate the mass fraction of this sample is expressed as Fe_3O_4.

(96.46%)

6-25 Weigh 0.1169g of pure NaCl, and dissolve it in water, then use K_2CrO_4 as an indicator, next add 20.00mL of $AgNO_3$ standard solution when titrating to find the concentration of the $AgNO_3$ solution. $M(NaCl)$=58.44 g·mol^{-1}.

(0.1000 mol·L^{-1})

6-26 Add 40.00mL 0.1020 mol·L^{-1} $AgNO_3$ solution to 25.00mL $BaCl_2$ solution, the remaining $AgNO_3$ solution needs to be back titrated with 15.00mL 0.09800 mol·L^{-1} NH_4SCN solution, what is the mass of $BaCl_2$ contained in 25.00mL $BaCl_2$ solution? $M(BaCl_2)$=208.24 g·mol^{-1}.

(0.272g)

Appendix

附 录

Appendix I Dissociation constants for acids

name	formula	K_a	pK_a
acetic	CH_3COOH	1.75×10^{-5}	4.75
		5.8×10^{-3}	2.24
arsenic	H_3AsO_4	1.10×10^{-7}	6.96
		3.2×10^{-12}	11.50
benzoic	C_6H_5COOH	6.28×10^{-5}	4.202
boric	H_3BO_3	5.81×10^{-10}	9.236
carbonic	H_2CO_3	4.45×10^{-7}	6.352
		4.69×10^{-11}	10.329
citric	CH_2COOH $\mid$ $C(OH)COOH$ $\mid$ CH_2COOH	7.44×10^{-4}	3.128
		1.73×10^{-5}	4.761
		4.02×10^{-7}	6.396
chromic	H_2CrO_4	1.8×10^{-1}	0.74
		3.2×10^{-7}	6.49
formic	$HCOOH$	1.80×10^{-4}	3.745
hydrocyanic	HCN	6.2×10^{-10}	9.21
hydrofluoric	HF	6.8×10^{-4}	3.17
hydrogen sulfide	H_2S	9.5×10^{-8}	7.02
		1.3×10^{-14}	13.9
oxalic	$HOOCCOOH$	5.60×10^{-2}	1.252
		5.42×10^{-5}	4.266
phenol	C_6H_5OH	1.05×10^{-10}	9.98
		7.11×10^{-3}	2.148
phosphoric	H_3PO_4	6.32×10^{-9}	7.199
		7.1×10^{-13}	12.15
phosphorous	H_3PO_3	3×10^{-2}	1.5
		1.62×10^{-7}	6.79

续表

name	formula	K_a	pK_a
phthalic	$C_6H_4(COOH)_2$	1.12×10^{-3}	2.95
		3.90×10^{-6}	5.408
salicylic	$C_6H_4(OH)COOH$	1.07×10^{-3}	2.97
		1.82×10^{-14}	13.74
sulfuric	H_2SO_4	1.02×10^{-2}	$1.99(pK_{a2})$
sulfurous	H_2SO_3	1.23×10^{-2}	1.91
		6.6×10^{-8}	7.18

Appendix II Dissociation constants for bases

name	formula	K_b	pK_b
ammonia	NH_3	1.75×10^{-5}	4.75
aniline	$C_6H_5NH_2$	4.6×10^{-10}	9.34
ethanolamine	$HOCH_2CH_2NH_2$	3.2×10^{-5}	4.50
ethylamine	$C_2H_5NH_2$	5.6×10^{-4}	3.25
ethylenediamine	$NH_2C_2H_4NH_2$	8.5×10^{-5}	4.07
		7.1×10^{-8}	7.15
hexamine	$(CH_2)_6N_4$	1.4×10^{-9}	8.85
hydrazine	NH_2NH_2	1.0×10^{-6}	6.0
hydroxylamine	$HONH_2$	9.1×10^{-9}	8.04
methylamine	CH_3NH_2	4.2×10^{-4}	3.38
pyridine	C_5H_5N	1.7×10^{-9}	8.77
triethanolamine	$(HOCH_2CH_2)_3N$	5.8×10^{-7}	6.24

Appendix III Solubility product constants (18℃)

compound	K_{sp}	temperature/℃	compound	K_{sp}	temperature/℃
$Al(OH)_3$	2×10^{-32}		$AgOH$	1.52×10^{-8}	25
$AgBrO_3$	5.77×10^{-5}	25	AgI	1.5×10^{-16}	25
$AgBr$	4.1×10^{-13}		Ag_2S	1.6×10^{-49}	
Ag_2CO_3	6.15×10^{-12}	25	$AgSCN$	4.9×10^{-13}	
$AgCl$	1.56×10^{-10}	25	$BaCO_3$	8.1×10^{-9}	25
Ag_2CrO_4	9×10^{-12}	25	$BaCrO_4$	1.6×10^{-10}	

续表

compound	K_{sp}	temperature/℃	compound	K_{sp}	temperature/℃
BaC_2O_4	1.6×10^{-7}		Hg_2I_2	1.2×10^{-28}	
$BaSO_4$	8.7×10^{-11}		$MgNH_4PO_4$	2.5×10^{-13}	25
$Bi(OH)_3$	4.0×10^{-31}		$MgCO_3$	2.6×10^{-5}	25
$Cr(OH)_3$	5.4×10^{-31}		MgF_2	7.1×10^{-9}	
CdS	3.6×10^{-29}		$Mg(OH)_2$	1.8×10^{-11}	
$CaCO_3$	8.7×10^{-9}	25	MgC_2O_4	8.57×10^{-5}	
CaF_2	3.4×10^{-11}		$Mn(OH)_2$	4.5×10^{-13}	
$CaC_2O_4 \cdot H_2O$	1.78×10^{-9}		MnS	1.4×10^{-15}	
$CaSO_4$	2.45×10^{-5}	25	$Ni(OH)_2$	6.5×10^{-18}	
CoS_α	4×10^{-21}		$PbCO_3$	3.3×10^{-14}	
CoS_β	2×10^{-25}		$PbCrO_4$	1.77×10^{-14}	
$CuIO_3$	1.4×10^{-7}	25	PbF_2	3.2×10^{-8}	
CuC_2O_4	2.87×10^{-8}	25	PbC_2O_4	2.74×10^{-11}	
CuS	8.5×10^{-46}		$Pb(OH)_2$	1.2×10^{-15}	
$CuBr$	4.15×10^{-8}	18~20	$PbSO_4$	1.06×10^{-8}	
$CuCl$	1.02×10^{-7}	18~20	PbS	3.4×10^{-28}	
CuI	1.1×10^{-12}	18~20	$SrCO_3$	1.6×10^{-9}	25
Cu_2S	2×10^{-47}	18~20	SrF_2	2.8×10^{-9}	
$CuSCN$	4.8×10^{-15}	25	SrC_2O_4	5.61×10^{-8}	
$Fe(OH)_3$	3.5×10^{-38}		$SrSO_4$	3.81×10^{-7}	
$Fe(OH)_2$	1.0×10^{-15}		$Sn(OH)_4$	1.0×10^{-57}	
FeC_2O_4	2.1×10^{-7}	25	$Sn(OH)_2$	3×10^{-27}	
FeS	3.7×10^{-19}		$TiO(OH)_2$	3×10^{-29}	
HgS	$4 \times 10^{-53} \sim 2 \times 10^{-49}$		$Zn(OH)_2$	1.2×10^{-17}	18~20
Hg_2Br_2	1.3×10^{-21}	25	ZnC_2O_4	1.35×10^{-9}	
Hg_2Cl_2	2×10^{-18}		ZnS	1.2×10^{-23}	

Appendix Ⅳ Stability constants of the complexes

matallic ion	ionic strength	n	$\lg \beta_n$
氨配合物			
Ag^+	0.1	1,2	3.40,7.40
Cd^{2+}	0.1	1,⋯,6	2.60,4.65,6.04,6.92,6.6,4.9
Co^{2+}	0.1	1,⋯,6	2.05,3.62,4.61,5.31,5.43,4.75
Cu^{2+}	2	1,⋯,4	4.13,7.61,10.48,12.59
Ni^{2+}	0.1	1,⋯,6	2.75,4.95,6.64,7.79,8.50,8.49
Zn^{2+}	0.1	1,⋯,4	2.27,4.61,7.01,9.06

matallic ion	ionic strength	n	$\lg\beta_n$
氟配合物			
Al^{3+}	0.53	1,…,6	6.1,11.15,15.0,17.7,19.4,19.7
Fe^{3+}	0.5	1,2,3	5.2,9.2,11.9
Th^{4+}	0.5	1,2,3	7.7,13.5,18.0
TiO^{2+}	3	1,…,4	5.4,9.8,13.7,17.4
Sn^{4+}	*	6	25
Zr^{4+}	2	1,2,3	8.8,16.1,21.9
氯配合物			
Ag^+	0.2	1,…,4	2.9,4.7,5.0,5.9
Hg^{2+}	0.5	1,…,4	6.7,13.2,14.1,15.1
碘配合物			
Cd^{2+}	*	1,…,4	2.4,3.4,5.0,6.15
Hg^{2+}	0.5	1,…,4	12.9,23.8,27.6,29.8
氰配合物			
Ag^+	0～0.3	1,…,4	—,21.1,21.8,20.7
Cd^{2+}	3	1,…,4	5.5,10.6,15.3,18.9
Cu^+	0	1,…,4	—,24.0,28.6,30.3
Fe^{2+}	0	6	35.4
Fe^{3+}	0	6	43.6
Hg^{2+}	0.1	1,…,4	18.0,34.7,38.5,41.5
Ni^{2+}	0.1	4	31.3
Zn^{2+}	0.1	4	16.7
硫氰酸配合物			
Fe^{3+}	*	1,…,5	2.3,4.2,5.6,6.4,6.4
Hg^{2+}	1	1,…,4	—,16.1,19.0,20.9
硫代硫酸配合物			
Ag^+	0	1,2	8.82,13.5
Hg^{2+}	0	1,2	29.86,32.26
柠檬酸配合物			
Al^{3+}	0.5	1	20.0
Cu^{2+}	0.5	1	18
Fe^{3+}	0.5	1	25
Ni^{2+}	0.5	1	14.3
Pb^{2+}	0.5	1	12.3
Zn^{2+}	0.5	1	11.4
磺基水杨酸配合物			
Al^{3+}	0.1	1,2,3	12.9,22.9,29.0
Fe^{3+}	3	1,2,3	14.4,25.2,32.2
乙酰丙酮配合物			
Al^{3+}	0.1	1,2,3	8.1,15.7,21.2
Cu^{2+}	0.1	1,2	7.8,14.3
Fe^{3+}	0.1	1,2,3	9.3,17.9,25.1

matallic ion	ionic strength	n	$\lg\beta_n$
邻二氮菲配合物			
Ag^+	0.1	1,2	5.02,12.07
Cd^{2+}	0.1	1,2,3	6.4,11.6,15.8
Co^{2+}	0.1	1,2,3	7.0,13.7,20.1
Cu^{2+}	0.1	1,2,3	9.1,15.8,21.0
Fe^{2+}	0.1	1,2,3	5.9,11.1,21.3
Hg^{2+}	0.1	1,2,3	—,19.65,23.35
Ni^{2+}	0.1	1,2,3	8.8,17.1,24.8
Zn^{2+}	0.1	1,2,3	6.4,12.15,17.0
乙二胺配合物			
Ag^+	0.1	1,2	4.7,7.7
Cd^{2+}	0.1	1,2	5.47,10.02
Cu^{2+}	0.1	1,2	10.55,19.60
Co^{2+}	0.1	1,2,3	5.89,10.72,13.82
Hg^{2+}	0.1	2	23.42
Ni^{2+}	0.1	1,2,3	7.66,14.06,18.59
Zn^{2+}	0.1	1,2,3	5.71,10.37,12.08

Appendix V Conditional electrode potentials

half-reaction	$\varphi^{\ominus\prime}$ /V	medium
$Ag^{2+}+e^- \rightleftharpoons Ag^+$	1.927	4 mol·L^{-1} HNO$_3$
$Ce^{4+}+e^- \rightleftharpoons Ce^{3+}$	1.70	1 mol·L^{-1} HClO$_4$
	1.61	1 mol·L^{-1} HNO$_3$
	1.44	0.5 mol·L^{-1} H$_2$SO$_4$
	1.28	1 mol·L^{-1} HCl
$Co^{3+}+e^- \rightleftharpoons Co^{2+}$	1.85	4 mol·L^{-1} HNO$_3$
$Co(ethylenediamine)_3^{3+}+e^- \rightleftharpoons Co(ethylenediamine)_3^{2+}$	−0.2	0.1 mol·L^{-1} KNO$_3$+0.1 mol·L^{-1} ethylenediamine
$Cr^{3+}+e^- \rightleftharpoons Cr^{2+}$	−0.40	5 mol·L^{-1} HCl
	1.00	1 mol·L^{-1} HCl
	1.025	1 mol·L^{-1} HClO$_4$
$Cr_2O_7^{2-}+14H^++6e^- \rightleftharpoons 2Cr^{3+}+7H_2O$	1.08	3 mol·L^{-1} HCl
	1.05	2 mol·L^{-1} HCl
	1.15	4 mol·L^{-1} H$_2$SO$_4$
$CrO_4^{2-}+2H_2O+3e^- \rightleftharpoons 2CrO_2^-+4OH^-$	−0.12	1 mol·L^{-1} NaOH
$Fe^{3+}+e^- \rightleftharpoons Fe^{2+}$	0.73	1 mol·L^{-1} HClO$_4$
	0.71	0.5 mol·L^{-1} HCl

half-reaction	$\varphi^{\ominus\prime}$ /V	medium
	0.68	1 mol·L^{-1}H$_2$SO$_4$
	0.68	1 mol·L^{-1}HCl
	0.46	2 mol·L^{-1}H$_3$PO$_4$
	0.51	1 mol·L^{-1}HCl+0.25 mol·L^{-1}H$_3$PO$_4$
H$_3$AsO$_4$+2H$^+$+2e$^-$ ⇌ H$_3$AsO$_3$+H$_2$O	0.557	1 mol·L^{-1}HCl
	0.557	1 mol·L^{-1}HClO$_4$
Fe(EDTA)$^-$+e$^-$ ⇌ Fe(EDTA)$^{2-}$	0.12	0.1 mol·L^{-1}EDTA, pH 4~6
Fe(CN)$_6^{3-}$+e$^-$ ⇌ Fe(CN)$_6^{4-}$	0.48	0.01 mol·L^{-1}HCl
	0.56	0.1 mol·L^{-1}HCl
	0.71	1 mol·L^{-1}HCl
	0.72	1 mol·L^{-1}HClO$_4$
I$_2$(solution)+2e$^-$ ⇌ 2I$^-$	0.628	1 mol·L^{-1}H$^+$
I$_3^-$+2e$^-$ ⇌ 3I$^-$	0.545	1 mol·L^{-1}H$^+$
MnO$_4^-$+8H$^+$+5e$^-$ ⇌ Mn^{2+}+4H$_2$O	1.45	1 mol·L^{-1}HClO$_4$
	1.27	8 mol·L^{-1}H$_3$PO$_4$
Os^{8+}+4e$^-$ ⇌ Os^{4+}	0.79	5 mol·L^{-1}HCl
SnCl$_6^{2-}$+2e$^-$ ⇌ SnCl$_4^{2-}$+2Cl$^-$	0.14	1 mol·L^{-1}HCl
Sn^{2+}+2e$^-$ ⇌ Sn	−0.16	1 mol·L^{-1}HClO$_4$
Sb^{5+}+2e$^-$ ⇌ Sb^{3+}	0.75	3.5 mol·L^{-1}HClO$_4$
Sb(OH)$_6^-$+2e$^-$ ⇌ SbO$_2^-$+2OH$^-$+2H$_2$O	−0.428	3 mol·L^{-1}NaOH
SbO$_2^-$+2H$_2$O+3e$^-$ ⇌ Sb+4OH$^-$	−0.675	10 mol·L^{-1}KOH
Ti^{4+}+e$^-$ ⇌ Ti^{3+}	−0.01	0.2 mol·L^{-1}H$_2$SO$_4$
	0.12	2 mol·L^{-1}H$_2$SO$_4$
	−0.04	1 mol·L^{-1}HCl
	−0.05	1 mol·L^{-1}H$_3$PO$_4$
Pb^{2+}+2e$^-$ ⇌ Pb	−0.32	1 mol·L^{-1}NaAc
	−0.14	1 mol·L^{-1}HClO$_4$
UO$_2^{2+}$+4H$^+$+2e$^-$ ⇌ U^{4+}+2H$_2$O	0.41	0.5 mol·L^{-1}H$_2$SO$_4$

Reference

参考文献

[1] 高职高专化学教材编写组 . 分析化学 . 第 4 版 . 北京：高等教育出版社，2014.

[2] 丁保君 . 分析化学（双语版）. 第 2 版 . 大连：大连理工大学出版社，2017.

[3] 高金波，吴红 . 分析化学 . 北京：中国医药科技出版社，2016.

[4] 王英健，尹兆明 . 无机与分析化学 . 北京：化学工业出版社，2018.

[5] 张丽 . 分析化学 . 北京：科学出版社，2017.

[6] 王嗣岑，朱军 . 分析化学 . 北京：科学出版社，2019.

[7] Daniel C. Harris. Quantitative Chemical Analysis, 9th Edition, W. H. Freeman and Company, 2015.

[8] 丁瑞芳，段煜 . 分析化学双语实验 . 北京：中国医药科学出版社，2019.

[9] David S. Hage, James D. Carr. 分析化学和定量分析（英文版）. 北京：机械工业出版社，2012.

[10] F. W. Fifield, D. Kealey. Principles and Practice of Analytical Chemistry, 5th Edition. Blackwell Science Ltd, 2000.